D1334344

College of Ripon & York St. John
3 8025 00415165 3

The Art of Detective Fiction

Also by Warren Chernaik

THE POETRY OF LIMITATION: A Study of Edmund Waller

THE POET'S TIME: Politics and Religion in the Work of Andrew Marvell

SEXUAL FREEDOM IN RESTORATION LITERATURE

* MODERNIST WRITERS AND THE MARKET PLACE (*co-editor*)

* MARVELL AND LIBERTY (*co-editor*)

TEXTUAL MONOPOLIES: Literary Copyright and the Public Domain (*co-editor*)

Also by Martin Swales

ARTHUR SCHNITZLER

THE GERMAN NOVELLE

THE GERMAN BILDUNGSROMAN

THOMAS MANN

GOETHE'S "WERTHER"

EPOCHENBUCH REALISMUS

Also by Robert Vilain

HUGO VON HOFMANNSTHAL AND FRENCH SYMBOLIST POETRY (*forthcoming*)

YVAN GOLL – CLAIRE GOLL: Texts and Contexts (*co-editor with Eric Robertson*)

* *From the same publishers*

The Art of Detective Fiction

Edited by
Warren Chernaik
Martin Swales
and
Robert Vilain

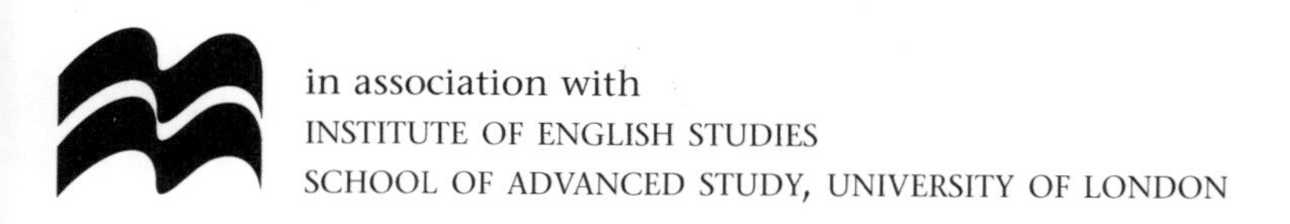

First published in Great Britain 2000 by
MACMILLAN PRESS LTD
Houndmills, Basingstoke, Hampshire RG21 6XS and London
Companies and representatives throughout the world

A catalogue record for this book is available from the British Library.

ISBN 0–333–74601–5

First published in the United States of America 2000 by
ST. MARTIN'S PRESS, INC.,
Scholarly and Reference Division,
175 Fifth Avenue, New York, N.Y. 10010

ISBN 0–312–22989–5

Library of Congress Cataloging-in-Publication Data
The art of detective fiction / edited by Warren Chernaik, Martin Swales, and Robert Vilain.
p. cm.
Includes bibliographical references and index.
ISBN 0–312–22989–5
1. Detective and mystery stories—History and criticism—Congresses.
I. Chernaik, Warren. II. Swales, Martin. III. Vilain, Robert.

PN3448.D4 A78 2000
809.3'872—dc21

99–046987

This book is printed on paper suitable for recycling and made from fully managed and sustained forest sources.

10 9 8 7 6 5 4 3 2 1
09 08 07 06 05 04 03 02 01 00

Printed and bound in Great Britain by
Antony Rowe Ltd, Chippenham, Wiltshire

Contents

Acknowledgements

The editors would like to thank the British Academy, Pro Helvetia and the School of Advanced Study, University of London, for grants in support of the international conference 'Murder in Bloomsbury'. The conference was held in June 1996 under the auspices of the School of Advanced Study, and with the collaboration of the University of London Institutes of English Studies, Germanic Studies and Romance Studies. We should also like to thank the many scholars from Australia, Austria, Canada, Denmark, India, Italy, Germany, Spain, Sweden, Switzerland, the United Kingdom and the United States of America who gave papers at the conference, especially John Bayley, Robert Barnard, Renate Böschenstein, Karl Guthke, Stephen Knight, Klaus Lüderssen and Morag Shiach. We should like to thank the novelist Liza Cody for her spirited and illuminating public interview with B.J. Rahn, one of the highspots of the conference, and Larry Beinhart for his masterful inaugural lecture. We are indebted also to Professor B.J. Rahn, Professor John Flood and Professor Annette Lavers for their advice and assistance, and to Rebecca Dawson, Karin Hellmer and Jane Lewin for their help in organizing the 1996 conference. Special thanks are due also to Charmian Hearne at Macmillan for her support and assistance in producing this volume of essays.

We are grateful to Capra Press for permission to include extracts from two books by Ross Macdonald (Kenneth Millar), *Of Crime Writing* and *Self-Portrait: Ceaselessly into the Past*; to *The Armchair Detective* for material from interviews published in that journal; and to Chatto and Windus for permission to include an excerpt from Philip Kerr, *A Philosophical Investigation*.

Notes on the Contributors

Birgitta Berglund is Lecturer in English at Lund University, Sweden. She is the author of *Woman's Whole Existence: the House as an Image in the Novels of Ann Radcliffe, Mary Wollstonecraft and Jane Austen* (1993). She is a member of the Swedish Detective Fiction Society and has lectured widely on women writers and the rôle of women in detective fiction.

Jonathan C. Brown is a British Academy Postdoctoral Fellow affiliated to the French Department at University College London. He did his first degree at Cambridge and went on to do a PhD at UCL. He is currently working on a book entitled *The French Writer-Filmmakers: from Intervention to Interplay*.

Warren Chernaik is Emeritus Professor of English at the University of London, Visiting Professor at the University of Southampton, and Senior Research Fellow, Institute of English Studies, School of Advanced Studies, University of London. His publications include *The Poetry of Limitation: a Study of Edmund Waller* (1968), *The Poet's Time: Politics and Religion in the Work of Andrew Marvell* (1983), and *Sexual Freedom in Restoration Literature* (1995). He is co-editor of two other volumes of essays published by Macmillan in association with the Institute of English Studies, *Modernist Writers and the Marketplace* (1996) and *Marvell and Liberty* (1999), and of *Textual Monopolies: Literary Copyright and the Public Domain* (1997).

Sarah Dunant is a novelist, cultural commentator and broadcaster. She was written eight crime thriller novels, including *Fatlands* (1993), *Under my Skin* (1995), *Transgressions* (1997), and the recent *Mapping the Edge* (1999), and has edited two books of essays, *The War of the Words* (1994), on the political correctness debate and (with Roy Porter) *The Age of Anxiety* (1995), on Millennial anxieties.

Patrick ffrench is Lecturer in French at University College London. He is the author of *The Time of Theory: a History of 'Tel Quel'* (1996) and of *The Cut: Reading George Bataille's 'Histoire de l'œil'* (1999), and the editor, with Roland-François Lack, of *The 'Tel Quel' Reader*. He is also the author of a number of articles on critical theory and on French literature.

David Glover is Lecturer in English at the University of Southampton. His most recent book is *Vampires, Mummies and Liberals: Bram Stoker and the Politics of Popular Fiction* (1996), and he has published widely on popular culture and cultural theory. He is Editor of the journal *New Formations*.

Margaret Kinsman is Senior Lecturer in English at South Bank University. She is the author of articles on crime fiction and mystery-writer profiles in the *St James Guide to Crime and Mystery Writers, Women Times Three, Mystery and Suspense Writers* in the Scribners Writers Series, and the forthcoming *Oxford Companion to Crime and Mystery Writers* and *Diversity and Detective Fiction.* She is a member of the Popular Crime Association's crime and mystery fiction caucus and the British Crime Writer's Association. A longer work on Sara Paretsky is in progress.

Audrey Laski is retired from running the Education Department at the Central School of Speech and Drama. She has published six novels, none of them a detective story, and is working on a book about serial detectives.

Laura Marcus is Reader in English at the University of Sussex. Her publications include *Auto/biographical Discourses* (1994) and *Virginia Woolf* (1997). She has also edited *Twelve Women Detective Stories* (1997) and Marie Belloc Lowndes, *The Lodger* (1996). She is currently working on *The Tenth Muse: Cinema and Modernism.*

Peter Messent is Reader in Modern American Literature at the University of Nottingham. He is the author of *New Readings in the American Novel* (1990, 1998) and *Mark Twain* (1997), and editor of *Criminal Proceedings: the Contemporary American Crime Novel* (1997). He is working on a study of the short fiction of Mark Twain.

Martin Priestman is Reader in English at Roehampton Institute London. He was educated at the Universities of Cambridge and Sheffield, and his publications include *Cowper's 'Task': Structure and Influence* (1983), *Detective Fiction and Literature: the Figure on the Carpet* (1990), and *Crime Fiction from Poe to the Present* (1998). He has a book forthcoming, *Romantic Atheism: Poetry and Freethought, 1780–1830,* and has been commissioned to edit *The Cambridge Companion to Crime Fiction.*

David Schmid is Assistant Professor in the Department of English at the State University of New York at Buffalo, where he teaches classes in twentieth-century British and American fiction, popular culture and cultural studies. He is currently at work on two projects: *Mad, Bad and Dangerous to Know: Serial Murder and Contemporary American Culture* and *Mean Streets: Space in Detective Fiction.*

Josef Škvorecký emigrated to Canada after the Soviet invasion of Czechoslovakia in 1968. From 1971 he was Professor of English at Erindale College, University of Toronto, as well as keeping Czech literature alive through 68 Publishers, the Czech-language publishing house run with his wife, Zdena Salivarova. His major novels include *The Bass Saxophone* (1977), *The Miracle Game* (1972, 1991) and *The Engineer of Human Souls* (1985). His detective stories include *The Mournful Demeanour of Lieutenant Boruvka* (1988), *The Return of Lieutenant Boruvka* (1990), *The End of Lieutenant Boruvka* (1990) and *Sins for Father Knox* (1989).

Martin Swales was educated at the universities of Cambridge and Birmingham. He currently holds the Chair of German at University College London. He has written monographs on Arthur Schnitzler (1971), the German *Novelle* (1977), the German *Bildungsroman* (1998), Thomas Mann (1980), Goethe's *Werther* (1987), and German realistic fiction (1995 and 1997).

David Trotter is Quain Professor and Head of the English Department at University College London. He is the author of, among other books, *The English Novel in History, 1895–1920* (1993), and, with Sandra Kemp and Charlotte Mitchell, *Edwardian Fiction: an Oxford Companion* (1997). His book *Cooking with Mud: the Idea of Mess in Nineteenth-Century Art and Fiction* is forthcoming.

Robert Vilain was educated at the University of Oxford and is Lecturer in German at Royal Holloway, University of London. He has written a monograph on Hugo von Hofmannsthal (forthcoming), and articles on European literature in the nineteenth and twentieth centuries, especially its relations with music. He has co-edited a volume of essays on Yvan and Claire Goll (1997) and is preparing a critical study and (with Andreas Kramer) a bibliography of the works of Yvan Goll.

Chris Willis teaches at Birkbeck College, University of London. She worked with Laura Marcus on *Twelve Women Detective Stories* (1997), has published several articles on popular fiction and is co-editor of *The New Woman in Fiction and in Fact,* to be published in association with the Institute of English Studies (2000).

Introduction

Martin Swales

The essays collected in this volume began life as papers delivered at an international conference on the detective story, which was held in London, under the auspices of the University of London's School of Advanced Study, in June 1996. That conference was the brainchild of three research institutions of London University – the Centre for English Studies, the Institute of Germanic Studies and the Institute of Romance Studies. The upshot of this collaboration was, as can readily be imagined, a very international conference. In the Acknowledgments to this volume we pay tribute to the contributors and a number of helpers without whose creative input the whole project would not have got off the ground. When we came to publish the volume, it became clear that simple considerations of space would decree that only a selection of the papers delivered could be incorporated in the final product. It cannot be too strongly stressed, however, that the whole was greater than the sum of the parts, and that some of that greater whole can still be felt as an animating presence behind the selection of papers that appear in this volume. Our title speaks of the 'art' of detective fiction, and questions of artistic statement are very much to the fore. But that 'art' is anything but hermetic, and constantly issues of broader (moral, legal, cognitive) import make themselves heard – as they should, and as they did with splendid vigour at the conference. As the conference showed, the detective story is a genre which attracts the love and attention of both academic and lay readers alike. It is our hope that this volume of essays will have a similarly broad appeal – and will, at the very least, bear witness to the fact that the detective story is a profoundly exciting genre.

That excitement is fuelled by a broad range of concerns which extend all the way from (basic) entertainment to high art. At one level, the detective story is the very paradigm of the 'rattling good story'; the reader cannot put the book down (as the saying goes) because of the sheer compulsion to find the explanation of 'whodunit'. At quite another level, the detective story enshrines that perennial hermeneutic of the narrative mode which works with deep structures of change and stasis, of onward-moving events and retrospective reflection, of mystery and its resolution. Within the simple and/or profound terms of plot-driven interest and excitement, then, the detective story is a genre of surpassing durability. But, as we all know, plots have a tendency to thicken; events are not enough to account for the fascination of the detective story. More is at stake. What, then, is at the heart of the genre's appeal?

The answer – as the papers in this volume most vigorously and variously suggest – has to do with a persistent dialectic which quickens the genre at every turn. It is the dialectic inherent in a tale which, of its very nature, speaks of mayhem – of criminality, transgression, violence, carnage, and (most usually) death – but which at the same time houses this Dionysian tumult and chaos within parameters of Apolline order, of (social and literary) convention, of rational explanation. In consequence, the genre can be at one and the same time conservative and subversive (see Glover, Chernaik and Messent). The detective story can validate the status quo ultimately as proof against the depredations of chaos and carnage, but it can also show that that victory of order over mayhem is Pyrrhic, one achieved by luck rather than judgement, one that is sustained less by drums and trumpets than by a sense of skin-of-the-teethdom. The explanation of the mystery may offer only limited exorcism of the anarchy that is so abundantly acknowledged; solution is not the same as resolution; closure may be temporary rather than permanent. Within this dialectic, all manner of interpretative and conceptual inflections are possible.

One is, so to speak, theological. A number of commentators have suggested that the detective story addresses our inborn sense of guilt. The inroads of criminality merely confirm what we have always known – that we are sinful creatures, prone to choose the darkness rather than the light. Yet the detective story does also hold out the promise of solace and uplift. It brings light into dark places, and, in so doing, for a brief period at any rate, it washes the world clean, it makes the prelapsarian condition possible. Not unrelated to this theological inflection of the detective story genre is the concern that one might call thanatological. Death constantly stalks the detective story; and it can be brutally acknowledged as the dismantling of body, mind, personality, selfhood, as cessation, eradication, silence. Yet – and here once again the dialectic is wonderfully beguiling – the body is an all-important clue, a material trace; very often the inert materiality of the corpse can come alive, can speak, can be an instance of judgement. Carnage and mess resolve themselves into clarity and articulacy, muteness gives way to eloquence (as Dunant and Trotter suggest).

Other forms of the dialectic are cognitive-cum-philosophical, as rationality and logic do battle with the irrationality of psychological promptings. From this aspect, one form of criminal comes particularly into his (and it is usually his) own – and that is the serial killer. Seriality is inseparable from the excitement of the detective story, as Priestman and Schmid indicate. At the simplest level, it imparts the urgency of the chase: the murderer has to be caught before he can kill again. Hence, the retrospective force of the analysis of past events acquires an urgent, forward-directed thrust. At more complex levels, seriality engenders a sense of pattern while withholding the explanation of that pattern. What seems random to the quotidian mind probably has coherence for the pathological mind. Familiar motivations for murder – greed, ambition, jealousy – no longer prove adequate to account for the incalculable in repetitive form. Reason is, in every sense, brought to its limits – and to its knees – by the force of unregenerate, frequently sexual, obsession.

The sexual charge behind much detective story writing has been accentuated rather than diminished by the presence within the tradition of key women writers. Their voice has a peculiar urgency when it engages with violence, which is by no means exclusively, but is very largely, a male province. (Dunant, Berglund and Kinsman explore these questions.) What is at issue is not just the portrayal of violence as such, but the cognitive and affective shock waves which it causes. Of late, women writers have created female detective figures who answer the male dialectic of the hyper-rational on the one hand and the hyper-violent on the other by the female dialectic of sophisticated – and often highly intuitive – thinking on the one hand and physical-cum-emotional vulnerability on the other. In any event, both in the classic 'golden age' mode and in its more modern forms, the detective story has been immeasurably enriched by the presence of women writers and by a thematics of the feminine.

Both male and female writers have been particularly adept at what one might term the cultural version of the governing dialectic of the detective story. This has to do with the genre's unmistakable fondness for a setting which is not just an enclosed world but also a world of high sophistication. At one level, this generates a particularly resonant theme which interlocks with a (more or less Freudian) perception of civilization and its discontents. High culture, so often taken to be the domain of the superego, is seen to be powerless before the subversive energies of the id. Civilization may house barbarity within its walls. The urbanity of the bishop, the librarian, the professor may be the mask of savagery. At another level, the civilized setting compounds the intellectual pleasure of the genre; sophisticated societies, particularly mini-societies (the monastery, the university) tend to be semiotically complex. The detective, and we the readers, are confronted by a plethora of signs, enshrined in the complex civilities of human behaviour within a relatively enclosed, well-educated institution. Invariably some of the civilities and signs will be sinister. But, at first sight they may be indistinguishable from the communal signs, hidden like a needle in a haystack. Except that, in this case, the needle and the haystack are made of the same medium. The semiotic overkill can truly be deadly.

That deadliness is inseparable from a context of intense textuality; and this brings me to my final category of the dialectic – which is, in the broadest sense of the term, textual. I have already had occasion to invoke a Freudian model. Many commentators have registered the similarity between the Freudian case-history and the detective story (as ffrench's chapter shows). Common to both is the impetus to read the signs, to transmute the mystery of inscrutable behaviour into some kind of intelligible causality, to bring light into dark places. But the Freudian model of textuality is not the only one in operation. Others can come into play in order to articulate the commerce between surface and depth. The surface consists of a profusion of data. But which of these particularities counts? Which bits of matter matter? When does a fact become a clue? The answer may be relatively obvious when a dead body is suddenly discovered. But other clues are hidden within the teeming profusion of everyday materiality. It is central to the logic and the charm of the genre that profusion must yield to textual coherence. The interplay of opaqueness and intelligibility is, once again, at the dialectical heart of the detective story, as Brown and Marcus indicate.

Many detective stories are particularly memorable for their articulation of that complex chemistry, part battle, part love-affair, which binds the detective to the world through which he moves. In the hands of many practitioners, the genre acquires a generosity of atmospheric statement, a kind of (Barthesian) 'effect of the real', the upshot of which is to create an impression of truthfulness which goes far beyond the genre's key signature of puzzle-and-solution, of mystery-and-clarification. One key thematic aspect is the social configuration of the detective story world. Particularly within the American, hard-boiled tradition, the urban environment of the mean streets is the objective correlative and persuasive metaphor for the referential truth of the tale (see Chernaik's and Messent's discussion of these issues). The atmosphere of such narratives is anything but mere background; rather, it is central to the statement they make. And it is in that urban atmosphere (one that can be both realistic and surreal, as Vilain's contribution reminds us) that the detective figure lives, moves, and has his being. Moreover, in respect of the detective figure himself or herself, we are frequently confronted with a density of statement that makes the figure far more than simply the adept solver of puzzles. Often the detective is interesting by virtue of his or her relationships (or lack of them). In many examples the detective is strangely celibate, seeking to keep the tainting world at arm's length. But often there is a friend, a helper, a trusted sidekick. Gender rôles can be crucially at issue here. Frequently, the detective figure may exude a kind of tragic aura. Many of the great detectives are figures who are forced to handle the mayhem not only metaphorically (intellectually) but also literally (materially). That is to say: they exist in threatening proximity to the swamp of criminality and aberration. And in this sense, their knowingness may produce not superiority but rather manichean melancholy, a sense of weary complicity. They are perilously close to the abjection which they perceive as the all-pervasive condition of the human species around them. (Laski offers suggestive pointers to a taxonomy of the detective figure.)

In this sense, it is perhaps not too fanciful to claim that the detective may be the last tragic protagonist still available to our culture. The hero or heroine of tragedy in the grand manner not only suffers; that suffering produces knowledge, perhaps even insight. In the person of the detective, the insight coexists with a will to action; the detective understands the puzzle and is instrumental in bringing the perpetrators to justice. Insight does not, then, have to paralyse the will. Such heroism has a long lineage to it. We remember that Oedipus was the first tragic detective of our culture. Thebes, too, had its mean streets.

1
Poe and the Beautiful Segar Girl

Josef Škvorecký

There have already been articles and essays written under the title of this chapter, referring to Poe's story 'The Mystery of Marie Roget', and its source and background, the unsolved murder of Mary Rogers of New York. At one time, this unfortunate young woman made a living as a cigar or – using the spelling of the time – 'segar' girl, a shop attendant in John Anderson's cigar store on Broadway. But my title derives from the title of a feature film, *Poe and the Death of a Beautiful Girl*, which I wrote for the Czech Television Company. It was produced in the spring of 1996 in Prague, and edited and mixed at the Barrandov studios in that city. The writing of the script for the film was a labour of love, not a commercial enterprise. It was the result of a life-long affection for the first of the world's *poètes maudits*.

My admiration, and later love, for Poe began in early childhood; it originated in one of those paradoxical situations children find themselves in; in those times when they still respected their parents, and their parents loved them. I, too, respected my mom and dad, but, at the age of nine, I was also a secret sinner. Every day at 8 p.m. I went to bed as ordered. But as soon as my mother switched off the light in my bedroom, I pulled the blanket over my head, from under the mattress I removed a flashlight and a book, and began my nocturnal reading. Many children in my old country were guilty of this petty crime and I am told that it was not different in this country.

On the particular occasion I remember to this day, I 'borrowed' a book from the back row in my father's private library, where he kept works not thought suitable for children. It was *The Narrative of Arthur Gordon Pym of Nantuckett* and it scared me to death. The butchery on board the *Grampus*, the putrefying crew of the *Ship of the Dead*, the cannibalism performed on the poor murderer, Peters, and finally the black horrors of Tsalal Island drove me almost insane, if children can be driven insane. Naturally, at my age, I was unable to understand the difference between reality and imagination, and experienced *Pym* as so many early readers had: as reality. Being a child, I did not even suspect the author of exaggeration. The large shrouded figure of an alabaster-skinned person, emerging from the milky ocean, proved to be too much for me. That night I finished reading *Pym* and fell asleep with the book in my left hand and the flashlight in my right. The flashlight burned out over the night, but in the morning when my mother discovered me still sleeping with an open book in my hand, the burnt-out flashlight told its story. I was severely punished, but my father acknowledged my literary passion and provided another book for me. It had been recommended to him by the local bookseller as extremely suitable for schoolchildren: it was called *The Adventures of Huckleberry Finn*. My misinformed dad made a mistake. But these two books, extremely inappropriate as children's reading, hooked me on to American Literature.

In the course of a year I read all of Poe's stories and poems available to me; when I matured, even *Eureka*. I had not read most of his letters because they were inaccessible in my old country, though I read some when I made it to Charles University. In my devotion to the Master, I approached the zeal of his first French translator, Charles Baudelaire, except that I did not pray to Poe. In my teens I became fascinated by Poe's poetry and knew quite a few of his pieces by heart.

Now I must put this in context: I read all these works in Czech translation since I did not learn English until the advent of World War II: from the BBC Continental Service. I think you will agree with me that the Americans, and the British for that matter, are not nations of translators, although the situation has improved in recent years. The Czechs are. They almost lost their ancient language during the centuries of Germanization following the defeat of the Czech Protestant armies in the Battle of the White Hill near Prague in 1620. Toward the end of the eighteenth century, with nationalist philosophies emanating from Germany, many Czech intellectuals whose first language, by that time, was German – some did not even speak Czech – became aware of the fact that they, after all, were not one of the German-speaking nations, and devoted themselves to bringing the nearly dead language back to life. Since there was, at that time, very little contemporary Czech literature extant, only medieval literature – and that was written in a language about as comprehensible as Chaucer's English – they translated both Western classics and contemporary authors, so that by the middle of the nineteenth century all of Shakespeare was available in Czech, and Dickens's novels used to appear in translation only a couple of years after they had been published in England. Significant national Czech literature emerged only after a considerable body of world classics had been rendered into the language of the westernmost Slavs. Translating and reading translations developed into a habit, and consequently there are now many more translators *per capita* in the Czech Republic than in any of the great Western nations. Naturally, this means competition and leads to very high standards: so I was exposed to both the prose and the poetry of Edgar Allan Poe in a language that showed none of the awkwardness contemporary readers of Poe's originals complain about. Something similar happened to me as had happened to the French contemporaries of Poe, who were reading his works in Baudelaire's renditions.

Much later, when I became a Professor of English in Canada, I made good use of the richness of North American University libraries and familiarized myself with the many theories put forward by my colleagues, which concern every possible aspect of Poe's œuvre. I became a subscriber to *Poe Studies*, a scholarly magazine specializing in Poe, and an avid reader of various biographical books on the Master. One thing this obsession did not do: it never diminished my love for Poe.

Then, with a whimper, in 1989, communism ended in my native land, and from a non-person, which I had been for 20 years, I metamorphosed back into being a popular Czech writer. I went on a triumphant tour of the old country, which was like a dream come true: only the proverbial White Horse was missing. I was flooded with offers from publishers and film makers, with requests for articles from magazine editors, and I also received one proposal from a lady who worked as the dramaturge for the Czech Television Company. Her name was Helena Slavíková. She had read my essay on Poe which I wrote for the magazine *World Literature*, where I had been Editor before I became a non-person, and she asked me if it would be possible to write a feature-film script for her television company, based on 'The Mystery of Marie Roget', one of the subjects of my essay.

It was not only a challenge, but also the potential fulfilment of another dream of mine. I loved Poe, and I loved the movies: in fact, I might say that I grew up in the movie theatre of which my father was manager. So I immersed myself in the project suggested by a woman who, in the course of the work, proved to be not only an excellent editor, but also became my friend.

Let me now refresh your memory of 'Marie Roget'. The story is not one of Poe's best, but it is almost a truism among film makers that the better the literary story, the worse the film, give or take a few exceptions. Poe based his tale on an actual murder case, that of the cigar girl Mary Rogers, who was murdered and thrown into the Hudson River. A few days later, her lover, Danny Payne, committed suicide. For fear of libel, Poe transplanted the story to Paris and changed the names. Naturally, everybody in New York knew that Roget was Rogers.

The New York police proved unable to solve the case, but for several months the affair was intensively followed by the daily press. At that time Poe was dreaming of a magazine of his own, and saw in the girl's tragedy an opportunity for himself. If he could crack the case, it would bring him tremendous publicity and he might find a moneybag to finance his project. He naturally could not meddle with the efforts of the police, as fictional detectives – including his own – are in the habit of doing, and so he conceived an idea which gave birth to one of the seminal techniques of the detective story, later imitated by countless authors: the narrative of an Armchair Detective who bases his solutions exclusively on newspaper stories and never bothers to visit the scene of the crime in person.

For more than a century it was believed, especially by people who had only a second-hand knowledge of Poe's story, that he actually identified the murderer. Then came one of those ingenious American professors by the name of John Walsh, who convincingly – and breathtakingly – demonstrated that Poe solved nothing, and moreover that he panicked after the first two instalments had been printed in *Ladies' Companion*.[1] At the time the second instalment appeared, an abortionist, Mrs Loss, confessed on her deathbed that an unidentified medical doctor had performed an abortion on the cigar girl in the tavern she owned in Hoboken, which went wrong, and the girl died. In the first two instalments Poe clearly indicated that Mary's death was the result of murder, and he even, in general terms, identified the perpetrator. And now, instead of murder – bungled feticide.

Walsh noticed that between the second and third monthly instalments of the story in *Ladies' Companion*, there was an hiatus of one month. In the issue following the second part, the editor printed a story by Mrs Ellet which was almost exactly the length of the third instalment, printed one month later. From this, and from Poe's movements at that time, Walsh deduced that the frantic author persuaded editor Snowden to postpone the third and final instalment, in the hope that, somehow, he would be able to get new information on the spot – thus potentially abandoning a purely armchair solution – and crack the case. Much was at stake for him; his reputation as an intellectual problem-solver, and the dream of his own magazine.

However, not even Walsh offered an answer to the question of who killed Mary Rogers. Numerous serious solutions by literary scholars, and even more numerous pseudo-solutions by sensationalist writers were suggested to the public over the years; in at least one of them Poe himself qualified as the assassin. In 1971, three years after Walsh, Ray Paul came up with an explanation of the murder which I found very plausible and quite inspiring.[2]

The police in Poe's time believed that Mary died in Mrs Loss's establishment, of a botched abortion, on Sunday 25 July 1841. Paul argued that the abortion was, in fact, *successful*, that Mary recuperated in Mrs Loss's apartment on Monday 26 July and later that day was strangled to death by her lover Danny Payne, who, a couple of days later, killed himself by taking an overdose of laudanum.

Not a bad solution. However, there was one thing that had always intrigued me about the story. In his report, Poe's coroner asserts that Mary was a young woman 'of virtuous character'. I tried to find out whether this phrase, in the early nineteenth century, could have meanings other than the obvious one, but it seems that, in the coroner's terminology, it could signify only that one thing: Mary Rogers, prior to her rape by one or several blackguards, was a virgin. In his revision of the story, Poe let this finding of the coroner stand, but he also included the later confession of Mrs Loss concerning Mary's abortion. Unless we accept an immaculate conception, how could this be? In his final revision, before he included the story in his *Tales* (1845), he made quite a few cuts, additions and alterations, but he *overlooked* this discrepancy. He was known to be very meticulous about his corrections – so how could he overlook a blunder like this? But what other explanation could there be *except* blunder? And yet another problem: how was the coroner able to establish the *virginity* of a *raped* woman? I consulted a forensic pathologist, and it proved as much an enigma for him as it had been for me.

Yet another puzzle: the real-life coroner, Dr Richard Cook, in his report which describes Mary's mangled body in gruesome detail, states that he found *no traces of pregnancy*; moreover, he states this not just once, but *twice*. That certainly does not rhyme with abortion, which, after Mrs Loss's confession, was universally accepted.

Although I believe I read everything in print on Poe's story, I found this discrepancy mentioned only in one article.[3] The discrepancy is just noticed; no explanation – except, possibly, Poe's oversight – is offered. So the question remains: was Dr Cook correct and there was no abortion; or was he a blunderer who missed what must have been obvious about Mary's physical state at the time of her death? In the former case, the logical answer would be that Mary was murdered; in the latter case, the coroner's rightness as well as his blunder would both open possibilities.

I dismissed the idea that Dr Cook was right: I would have to follow the path of many solvers before me and that was not a prospect that would appeal to my imagination. Therefore I concentrated on the question of Dr Cook's expertise. First, I tried to find his report in the morgues of New York and New Jersey, but apparently it does not exist any more. Crucial to my deliberations on the doctor's competence or incompetence, was the question whether, in 1840, coroners did or did not perform an autopsy on the victims of murder. In this respect I consulted a man with a most fitting name, Mr Malvin Vitriol, Medical Librarian of Milton Helpern Library at the Institute of Forensic Medicine in New York. He confirmed that autopsies were performed in the early 1840s, but not routinely, just occasionally.

The question now was: how could Dr Cook assert with such certainty that Mary was not pregnant? If she was with child, the fœtus was still too small to show on the bruised body of the cigar girl. After a merely external examination, the coroner would not be able to tell with any certainty whether Mary was gravid or not. Yet Dr Cook, in his report printed in the daily papers, asserted *twice* that pregnancy was out of the question.

This may indicate that Dr Cook did what was very rare, but not impossible in those days: he did perform an autopsy. Is it probable, then, that he was such a nincompoop as not to see what had been going on in Mary's uterus? According to what I was able to establish, Dr Cook was an experienced coroner, not a beginner. And again, he stressed the non-pregnant state of the victim by repeating it twice. Therefore, if we exclude gross incompetence, there is only one answer: Dr Cook falsified his report on purpose.

But what was the purpose? In the Mary Rogers affair there was one shadowy figure, her employer John Anderson. Many years later he stated to a friend that Mary had been the indirect cause of his downfall as a budding politician in New York City. Before her tragic end, Mary had already disappeared once, and was absent from Anderson's store for several days. Notices about the disappearance and reappearance were printed in the New York papers at that time, and their tone was ironic: it hinted at something improper in her employer's behaviour. After Mrs Loss's disclosure, the press remembered the earlier affair, and unequivocally suggested that the story had repeated itself, only this time without a happy ending. Anderson was taken in for questioning but exonerated of any suspicion. Perhaps this was on the strength of the coroner's report?

At the time of the murder, Anderson was a married man with children, and a wealthy, if not already very rich man, as he certainly became later in life. He was a respected citizen and had political aspirations for a career in Tammany Hall. He could afford many things, but one he could not was political scandal. And that would certainly follow had he been identified as the man who twice made Mary pregnant and twice paid for her abortion. Influential public opinion believed abortion to be a heinous crime. A strong suspicion that he had committed such a crime would certainly have ruined Anderson's political hopes and would have caused trouble in his married life. True, gentlemen were tacitly permitted to have mistresses, but a tacit acceptance of a heinous crime was, I believe, too much to expect.

However, Anderson could afford to pay for many things, including a forged document. So I assumed that Dr Cook was in dire need of money, and Anderson supplied that money. Naturally, I would be hard pressed to substantiate this hypothesis, but I was writing a film script, not a scholarly treatise, and as we all know, after you eliminate everything that is impossible, whatever remains, however improbable, is – or should I say *might be* – the truth. Anyway, my solution to this problem in Poe's story was less improbable than, for instance, making Poe the assassin. Dr Cook, let us say, wanted to build a private sanatorium but his funds were inadequate. So he accepted Anderson's help in exchange for a rewriting of his autopsy report.

Now I had three elements on which to build the dramatic structure of my script: Poe's reasons for the hiatus between the two instalments, as explained by John Walsh; Ray Paul's hypothesis that Mary, after a successful abortion, was murdered on Monday by her beau Danny Payne; and my own imagined reason for Dr Cook's unbelievable incompetence. I could now start thinking about the form, the shape and the tone of the story.

Of Poe's actual tale I left some quotations in the voice-over and I made the story the subject of conversations, both in Poe's household and in Anderson's cigar store. In Poe's time, cigar stores such as Anderson's functioned to a degree like cafés in Europe: they were meeting places for sportsmen, writers and literary critics. Anderson's store, for instance, was frequented by, among others, Fitz-Greene Halleck, Nathaniel Parker Willis, James Fenimore Cooper and Poe himself. So rather than making much use of the Marie Roget story, I concentrated on the life of her model, Mary Rogers, and on the effect of Poe's mishandled detective effort both on his public standing and on his family.

As for the family, I found it extremely difficult to write scenes that would avoid the usual clichés about the gloominess and misery of Poe's household and infuse the scenes with life. However, Poe's domestic life, with his sick wife and his constantly worried mother-in-law, was indeed rather gloomy and not very inspiring. Then, luckily, I hit upon the idea that helped me: Poe's womanizing.

Among the many women who are known to have admired Poe and to have been admired by him, one stands out: Fanny Osgood. She was an established poet, the author of several volumes of verse, and her relationship with Poe seems to have been particularly passionate. Whether it was purely platonic is hard to say: one little fact remains, that after Fanny became pregnant, the *affaire* abruptly ended. Fanny's husband, a painter, did a portrait of Poe which has survived. Apparently no saint himself, he tolerated his wife's well-advertised passion for Poe. After Poe's death, Fanny wrote one of the most intense, because it seems to me very personal, poetic obituaries, which contains the lines: 'He haunts me forever, I worship him yet; / O! idle endeavor, I cannot forget!' So I decided to include this lovely woman (her likeness was reprinted in many books about the poet) in my script, although she had nothing to do with Poe's immediate problem; the solving of Mary Rogers's murder. Speaking commercially, a pretty young woman enlivens even the dullest of scripts, and in my own, Fanny performs a special function.

It is not always easy to reconcile Poe's known frolicking with pretty poetesses with his deep love for his sick young wife. However, Mrs Slavíková gave me a solidly feminine hint: 'Well,' she said, 'he had this sick child-wife at home whom he loved and adored. But after all, he was a man, and, from time to time, he simply needed the company of non-angelic women.' I took the hint, and in my script Virginia herself recognizes this emotional dichotomy and explains it to Fanny. The confrontation of the two young women, over Poe's manuscript eulogy of Fanny written for *The Literati of New York City*, is one of the focal points of the film script. Thus the little drama of Poe's affair with Fanny, together with the large drama of poor Mary Rogers's assassination, and Poe's feverish and futile attempts to find the murderer and save the dream of his own magazine, hopefully helped me to overcome the dangers of flatness and tedium that afflict so many biographical films.

By basing my script on 'Marie Roget', I was able to follow the advice of the Master, who stressed the all-important presence in one's artefact of the death of a beautiful woman. It made it possible for me to come up with not one, but *two* such all-important girls, and perhaps *three*, if we include the fictional Marie Roget. The success of my script seems, therefore, three times assured.

Last, but not least, I wanted Poe's beautiful poetry to resound from the small screen, in a magnificent Czech translation. So I loaded my vehicle with as much poetry as I could, sometimes in the voice-over, sometimes from the mouths of the poet himself and of his friends and girlfriends. I hope that hearing the poetry on today's most influential medium, the viewers will, the morning after, flock to the nearest bookshop.

The film is in the can now. Hopefully, some British or American television company will buy it and you will have the opportunity to see it. Perhaps someone with influential connections in the mass media will put in a word for this tribute to the Master, which comes from a foreign land where, a century-and-a-half after his death, he is a household name.

Notes

1 John Walsh, *Poe the Detective: The Curious Circumstances Behind 'The Mystery of Marie Roget'* (New Brunswick: Rutgers University Press, 1968).
2 Raymond Paul, *Who Murdered Mary Rogers?* (Englewood Cliffs: Prentice Hall, 1971).
3 Martin Roth, 'Mysteries of "The Mystery of Marie Roget"', *Poe Studies*, 22.2 (December 1989), 27–34.

2
Body Language: a Study of Death and Gender in Crime Fiction

Sarah Dunant

When as a crime writer I find myself in front of live audiences, one of the first things I am always asked is why crime fiction is so popular. It's an interesting question, not least because the very people asking it – fans of the genre – are the ones who, in theory, ought to be able to answer it. After all, they buy and voraciously read the stuff. Yet when you put the question back to them they never can. Or at least not satisfactorily.

They might say they like the compulsive nature of the stories, the way that crime fiction hooks them so that they always have to read to the end. If you push them a little further, they might admit that, although literary fiction may be more profound or better written, it is often not as much fun to read, its narrative neither powerful nor grabbing enough. In contrast, they say, you always know where you are with a good thriller or crime novel.

So this gives us one immediate reason for crime fiction's continued success: it is good at story telling. And the reasons it is good at story telling, of course, is because it offers a mystery, a series of questions that have to be answered. And that is the other thing people say: they like the intellectual puzzles of the books, the way that they are asked to pit their wits against the detective. Many readers agree that as a form of fiction it can sometimes be both frightening and violent, but (and I think this is a very powerful 'but') almost always within a fundamentally safe framework – havoc, mayhem, immorality and cruelty may abound, but you will be returned to sanity at the end, the good guys will win, order will be restored. Of course that is not true of all crime novels, but in general that sense of reassurance – like that of a nightmare from which you will be awakened or released – is an important ingredient in crime fiction's broad popularity.

If you push this dialogue further and talk about the importance of death within the form, audiences will observe that crime fiction does bring a taboo topic into the open and that in doing so it allows people to feel things that they might not feel elsewhere. That is what I want to concentrate on for the first half of this chapter – the power of death and the body in crime fiction. What is it about death in this genre that is different from death in any other kind of fiction? And does that in some way help to explain its popularity?

It goes without saying that within literature there are all kinds of novels that talk about death, often much more deeply and philosophically than any crime novel does (partly because, as we have already ascertained, a good crime novel needs to keep a narrative pace, which makes any sustained introspection or reflection a potentially dangerous meander). But what I would argue is that only crime fiction deals with death in a routine, even formulaic and ultimately reassuring way.

Crime fiction, by its very nature, has an intimate relationship with the dead body. There must be a crime in order for there to be a fiction and the most serious and obvious crime is murder. So be it from the classic formulaic crossword puzzles of Agatha Christie and the Golden Age to the more psychologically penetrating world of P.D. James, the stylish hard-boiled fiction of Chandler or Hammett, or the racy, violent thrillers of everyone from Mickey Spillane to Thomas Harris and a host of contemporary imitators – there must be a corpse or corpses.

Now if you think about it – the corpse is a very powerful object. It contains within it many of the most fundamental questions and terrors that we have about death. First of all, what exactly is it that happens when someone dies? Often in real life when death takes place – a non-violent death – there is no obvious outward manifestation of the moment itself. Someone just stops, they cease to be, which in many cases makes it even more difficult to grasp the central awesome fact of death, the disintegration or the disappearance of personality, ego, spirit, whatever you call it. Death means not just the death of the body – which in some cases still looks almost alive or at least at peace – but the loss of the person inside it.

This in turn provokes even deeper questions. Why should it be this person, rather than any other, who has died? – that is, the appalling random nature of the act. Why do we have such terrible lack of control? And following that, who or what, if anything, does have control? This at its most basic returns to the ultimate 'Is there a God?' question, 'What is the meaning of death?' and so on. Huge, unanswerable questions, as we all know. But what I'd like to suggest here is that one of the reasons that crime fiction is so popular is that it does, in one way or another, offer answers to all of these questions and paradoxes.

Now I am not claiming that it has the right answers – we know that is impossible – or even that its answers are particularly profound. What I am claiming is that within a popular, accessible form, crime fiction again and again offers a kind of artificial construct which addresses these concerns in a way that makes the reader feel more comfortable, as if it had in some way answered them.

Let's take the body, for example. Although there are crime novels where death is designed to be indistinguishable from natural causes, they are the minority within the genre. The whole point of the corpse in crime is to show death as a result of violence. You cannot mistake a dead person for a live one in crime fiction. There are no nineteenth-century wasted consumptives filled with serene beauty. And so that line of awful subtlety and confusion as to how exactly does life leave the body is drawn for you. It leaves in the purple face of strangulation, of the clotted mess of a stab wound, or the leaking bullet hole in the chest. Or in the more recent Tarantino world, the body blown to smithereens. You know where you are with a corpse in crime fiction. The person you once knew has gone, is dead, life has been ripped away.

What that doesn't help with is the question of what happens to that personality. And as we know, for the living that is one of the most terrible things about death; the loss and missing and gradually fading away of what had been that person, both in material terms, of a body which now has no purpose but to decay, and, in more abstract emotional terms, of your memory. But in many ways the amazing thing about crime fiction is that almost the opposite journey takes place. Because in crime fiction death is the start of something, not the end. More often than not, the body in some shape or form begins the story. Without the body, without the murder, there would be no fiction. Thus death is the beginning of the investigation. And for any investigation to succeed you must know more about the victim. So far from being gradually forgotten, the process of almost reactivating the victim is crucial to the story. At its most basic the dead have to talk.

This takes place in two ways. First, in practical terms, very often in crime fiction the body itself does the talking. A study of the corpse is vital: exactly what happened to it? How was it robbed of life? These questions are not just idle medical curiosity – they are an important part of the plot. Anyone who has ever read a pathologist's report on a murder victim (and a lot of crime fiction writers have) will be struck by the sheer weight of time and expertise and close attention that the corpse receives. It is as if the corpse has an active rôle to play after it is lifeless. I would argue that this is even more important now – witness the rise and rise of the pathologist hero, or rather heroine (I shall be going on to talk more about Patricia Cornwell). In the hands of a good crime fiction writer, then, the body has a life after death. It is exposed, investigated, documented and examined. It is given careful, almost loving, attention and as a result yields secrets and provokes thoughts. Far from being forgotten, the dead person stays around on the slab as a main character, albeit without words. In a key way the body is not dead.

Second, on a more abstract level, the rôle of the traditional detective is to bring alive the personality of the victim. Because it is only by knowing more about the person who died – what they were like, their past, their relationships, their work, their personalities, their passions, whom they loved, whom they hated or were frightened of – that the detective can start to unravel the mystery of their death.

So one of the central functions of crime fiction, I would argue, is to take a dead person and in some way make them alive again. This I would say addresses and soothes in a very primitive way one of the primary fears about death – that is, the complete loss and anonymity of it. But it doesn't end there, because the driving force of any crime story is to find answers to two basic questions, the one leading to the other: why was this person – or these persons – killed? And the second, even bigger question: who was responsible? And in so far as these two questions are at the centre of our fear of death, the answering of them is also at the centre of any good piece of crime fiction.

The first question – 'why this person?' – is about the discovery of motive. Love, inheritance, insanity, the desire to protect a monastic community from discovering a book about comedy. It is all about the same thing – why this person was killed rather than any other. Indeed when it comes to the current favourite within crime fiction – the phenomenon of the serial killer – the sheer randomness of the victim is a very powerful reflection of that sense of randomness of death in our own lives. Death is really not your fault. You just happen to be in the wrong place at the wrong time. Like catching small pox or developing cancer, there is a terrible but irrefutable capriciousness to it.

Which brings us to the ultimate question. From why to whom? Whodunit? Because someone always has done it in crime fiction. Though death may seem mysterious and random and terrible, though it may masquerade as an avenging angel, as natural causes, or even natural disaster, that angel/devil/disaster has by the end of the book a name and an identity. There is someone to blame. At its most crude – in modern society where God is dead, instead you get the butler, or the lover, or even at its most dastardly, the madman. You might prefer God (a divinity can be seen as delivering benign killing or even justice) but at least you are not left with the terror of not knowing. The mystery of death is sorted out.

Now as I said earlier, these answers are hardly profound. On the contrary, they are banal. They are neither the meaning of life nor the meaning of death, but they are there every time. Because they are not profound, neither are they very lasting. Real devotees, devoted fans of crime (often the least discriminating of readers) devour crime fiction with the same insatiable appetite that others do Romantic Fiction, precisely because it is hooking into something in the subconscious, addressing anxieties which cannot finally be alleviated, only temporarily quietened. Another day, another death, another book. The fear is as repetitious as the reassurance.

Nevertheless, the fact that a form of reassurance is there, that answers are given, that the mystery is solved, does go some way to explaining why a form whose very *raison d'être* is violent death and its aftermath has grown so popular during an era when older reassurances about death – those of religion and God – have ceased to be meaningful or effective for many ordinary people. Crime fiction is, if you like, a genre whose time has come.

It is also a genre which has developed and expanded considerably over the last half-century and it is the nature of some of those developments that I want to address in the second half of this chapter. We stay with death and bodies, but now we divide them by gender. As in life we have seen a war between the sexes going on, so elements of the same conflict have been played out within crime fiction. Powerful forces have been unleashed, specifically to do with sexual violence. And the body – especially the female body – has become a cultural battlefield over which men and women are, albeit obliquely, facing and confronting each other in print. Here are two separate descriptions of the female corpse. Prepare yourself for the sight of death:

> 1. The depths cleared again. Something moved in them that was not a board. It rose slowly, with an infinitely careless languor, a long dark twisted something that rolled lazily in the water as it rose. It broke surface casually, lightly, without haste. I saw wool, sodden and black, a leather jerkin blacker than ink, a pair of slacks. I saw shoes and something that bulged nastily between the shoes and the

cuffs of the slacks. I saw a wave of dark blond hair straighten out in the water and hold still for a brief instant as if with a calculated effect, and then swirl into a tangle again.

The thing rolled over once more and an arm flapped up barely above the skin of the water, and the arm ended in a bloated hand that was the hand of a freak. Then the face came. A swollen pulpy gray white mass without features, without eyes, without mouth. A blotch of gray dough, a nightmare with human hair on it.

A heavy necklace of green stone showed on what had been a neck, half-embedded, large rough green stones with something that glittered joining them together.

Bill Chess held the handrail and his knuckles were polished bones.

'Muriel.' His voice said croakily. 'Sweet Christ. It's Muriel.'

2. The unfortunate victim, twenty five year old Mary Woolnoth, was found naked in the basement of the offices of the Mylae Shipping Company in Jermyn Street, where she had worked for three years as a receptionist, her face beaten in by a claw hammer. ...

The poor woman's red silk suspender belt was tied tight around her neck although by this stage she was already dead. A Simpson's store carrier bag was later pulled over the victim's head covering her ruined features from view. Possibly this took place prior to intercourse.

Using a Christian Dior Crimson Lake lipstick from the victim's handbag the killer wrote some four-letter abuse onto her bare thighs and stomach. Immediately above the pubic line was written the word 'FUCK', while on the underside of her thighs and buttocks was written the word 'SHIT'. Across each breast was written the word 'TIT'. Last of all the killer drew a happy smiling face in the white plastic carrier. I say last, because there was evidence of the lipstick having crumbled more during this drawing.

The unhappy victim's vagina contained traces of a latex-based spermicidal compound consistent with the killer wearing a condom prior to intercourse. He was no doubt mindful of the need to avoid DNA profiling. The said spermicidal compound is of a type most commonly used by homosexuals because of its greater strength. In past years we have also found that this is also the average rapist's favourite condom for the same reason.

Now it would be tempting to begin by asking the gender of these two writers. But it would be a trick question because in fact they are both men. One is very well known indeed, the other less so, though both are successful, talented writers and both approach the form with a level of sophistication and style. But what separates them is a time-gap of 50 years.

The first passage was from *The Lady in the Lake* by Raymond Chandler, and was a description of the eponymous dead heroine.[1] Now Chandler, despite being a white knight on the mean streets, was not above being nasty to women. (Anyone in any doubt only has to read Marlowe's treatment of Carmen Sternwood in *The Big Sleep*.) And Chandler knew as well as anybody else that women were often murdered for reasons to do with their gender. But when it comes to the handling of the female corpse as in a scene like this, what Chandler is interested in is something about the horror and disfigurement of death universal to both sexes, rather than any kind of exploitation of the specific female form. True there are touches of femininity: the poignant swirl of dark blond hair in the water, the necklace grotesquely embedded in the neck, but they serve rather as almost a philosophical reminder of the way death destroys all semblance of beauty, and certainly neither in this passage nor in any other does Chandler come near to what I would describe as the covert – or maybe overt – misogyny of that second piece, which is the opening of a book by Philip Kerr.[2]

Kerr is actually a clever and successful contemporary writer and this book, *A Philosophical Investigation* – a kind of *Silence of the Lambs* meets *Blade Runner* – is a thriller about a serial killer who reads Wittgenstein and cuts up women, both on and off his computer. It is witty, literate – a cut above a lot of the contemporary serial killing genre – and it tries hard to have its cake and eat it by having that extremely nasty description read by the female detective who goes on to head the murder investigation. But while the book has things about it that are unusual, what is not unusual is that opening, voyeuristic image of a female corpse sexually mutilated, and served up with a helping of spurious police/forensic jargon to make us feel a tad less guilty about reading it. Descriptions like this one – and I could have chosen a dozen more vicious and graphic examples – litter the more hard-boiled and even some soft-boiled crime fiction nowadays.

Of course, juxtaposing Kerr's passage with Chandler's one might simply say that times have changed. We live in a society where all forms of expression – be it literature, journalism, cinema, or television – have moved the goal posts of taste in terms of what is acceptable to show and tell. That is certainly true. Popular fiction in general is more explicit across the board. Just think of Mills and Boon sex these days.

But while the last few decades have seen all kinds of bodies fictionally undressed, their holes and bumps and secret places explored, when it comes to explicit sexual violence, I would argue that within crime fiction women have been particularly targeted for mutilation and depersonalisation.

So what about the next argument: that if this kind of crime takes place in our society crime fiction should tell it like it is? Murder is a disgusting business. Sexual serial murder even more so. At least these images of female bodies are honest. Well, the first thing to say is, yes, of course, disgusting crimes do take place and many of them are aimed at women. But statistically there are not nearly so many as one is led to believe from the groaning shelves of contemporary thrillers. Second, far from being honest, I would argue that descriptions like Philip Kerr's are open to accusations of dishonesty, because they have less to do with social realism and more to do with voyeurism. I would say that whether or not the writer is aware of it, such passages encourage a sense of titillation. When we look at Mary Woolnoth's body in print we feel not only her death, but also her degradation, humiliation and terror (all qualities essential to these kind of deaths). The corpse itself has a brutal lack of identity. You will notice that very often such corpses have no face. Who they are is not important. It is their body and what has been sexually done to it that is the point. (It is interesting to note that while the body of Muriel in the Chandler passage is also faceless, a pulpy mess, that is as much about the effects of nature and the water as the viciousness of her killer.)

That obliteration of identity in the Kerr passage is important in another way, too. The point about these female victims of serial killers is that they are two-a-penny. And all they have in common – i.e. what attracts the killer towards them – is, in basic terms, not their faces but their genitalia. That is what makes them victims. And that, incidentally, is what makes them so unsatisfactory to reactivate when it comes to trying to solve the crime. For the detective they are often just a pathetic list of names. In terms of the traditional crime novel they are not always given the care and attention of reconstruction; one follows on after the other too quickly for that. It is just their gender that counts. And it is the way that such corpses have become an almost casual part of the language of death within the hard-boiled crime genre that makes this kind of book unacceptable to an increasing number of women readers.

It is of course not coincidental that during the same years that have seen an explosion in images of sexual violence towards women we have also seen an explosion in feminism. It is as if, at some subliminal cultural level, what we have got here is action and reaction. The more visible and noisy women have become in our culture, the more they have become the target of fictional violence and sadism.

It is almost as if their very equality has made them fair game. It is interesting in this context to look back to Jack the Ripper, who was, after all, the first serial sexual killer on whom we have any coherent documentation. And here I draw on historian Judith Walkowitz's excellent book *City of Dreadful Delight,*[3] which shows how the media of the day suggested that Jack's killing of prostitutes might well have had something to do with the fact that more women were on the streets – all women, not just working class women, since the 1880s saw the beginning of the first wave of feminism, with women going around unchaperoned. Women's very visibility somehow tempted male violence. The implicit argument is that if women chose to leave the safety of their homes, they will, by definition, be more vulnerable to sexual violence. Almost that they deserve what they get.

This brings me to the last element of this chapter: the way in which women have fought back. Because, of course, contemporary women writers have not taken this latest fictional assault lying down. Broadly speaking, they have responded in two ways. First, in the 1980s feminism and crime fiction formed almost an arranged marriage to create a whole sub-genre of crime fiction in the arena of the private eye (always the more hard-boiled male end of the market and more American than English). An increasing number of women writers, some British, some American, some overtly feminist, some less so, got on their trench coats, emptied their fridges, put a whisky bottle in their filing cabinet and, hey presto, the female private eye was born. Sara Paretsky, Sue Grafton, Linda Barnes, Liza Cody, Val McDermid, Sarah Schulman, Mary Wings, Joan Smith: the list is long and rich. And if I now go on to talk about them with some sense of ownership it is also because I am part of that list, with my three private eye books, *Birthmarks, Fatlands* and *Under my Skin.*[4]

What we had in common as writers (apart from the obvious fact that we created central women characters who were heroes and not victims) is that by and large we all dealt with violence, and particularly violence towards women, differently from men. To start with we didn't revel in it. Often we didn't go into it at all. Our plots ranged wide and far, through corruption, business, the domestic, the international, the political, the personal. Although most of us were interested in the dynamics of sexual politics, that was not the same thing as sexual violence. As a writer myself I thought long and hard about any violence I used in my books. And when I did use it, I and many like me went out of our way to treat it seriously, to avoid the gratuitous, the voyeuristic. Or – to put it more crudely – we wrote it in a way which meant we didn't get off on it, so neither could the reader. Instead we used wit, irony, pacy plots, provocative ideas and sassy women to offer the reader (mostly female but not exclusively) a world where there were active powerful women with whom they could identify rather than a just a catalogue of faceless defiled corpses.

It was a good strategy and a popular, commercially successful one. But – and it is a big 'but' – it was not enough. The constraints we placed upon ourselves opened us up to accusations of romanticism. (When did you last read a female private eye novel where the female private eye got raped? In life it would almost certainly be a constant threat, but our politics of wish fulfilment made it hard for us to handle it.) And while we were busy creating new heroines, mythic women for a mythic form, the boys, by and large, were still cutting up vaginas and dumping body parts in black bags in dustbins.

This brings me to the second stage of women's writing. Because there is no doubt that over the last few years women have started to get their hands in the muck too. The numbers of books where women writers – often avowedly feminist writers – have got down and dirty when it comes to writing about sex and violence has increased. From Patricia Cornwell's bodies on the slab with the gruesome, sometimes sickening stories of how they got there, to books like Susanna Moore's infamous *In the Cut*, told from the point of view of the victim of a serial killer, to my own recent novel, *Transgressions*, which is the story of stalking and rape from the viewpoint of the stalked woman and charts the way in which she turns the tables on her aggressor.

And the question with which I want to end this chapter is to ask how far, by playing the boys at their own game, are we women successfully subverting and challenging the nasty thriller, or how far might we just be falling into the trap of perpetuating it? Here, I admit, I find myself on shaky ground. I think there are ways in which (despite her evident brilliance) Patricia Cornwell's books are part of the problem rather than the solution. Although broken mutilated female bodies are cared for by Kay Scarpetta's gentle but firm female hands (better to have slashed vaginas examined by a women pathologist than a man), the underlying right-wing politics of her creator's vision – the world is full of evil, and all good can do is keep the levels of filth in the sewers down – means that she is more often than not playing into our fears rather than genuinely helping us to transcend them. The bleak message behind Cornwell's fiction is: put another set of locks on the doors, girls, because however strong you are, the bad men are worse than you think. On the other hand, I would have to say that though Susanna Moore's *In the Cut* is a shocking book, I think it is also a brave, and I would have to add, in places, an erotic one. My own novel, *Transgressions,* found itself at the centre of a similar debate: was it exploitative or subversive to have a woman emerge from rape as the victor rather than the victim? I would say it was deliberately subversive – but then I would, wouldn't I? I set out to punch an imaginative hole through the fear that I and many women have – that we have almost been trained to have by the way our culture targets us – that at some point in our lives the psychopath will get into our house and terrorize us. And when that happens how would we cope? What would we do not to just survive but to triumph?

My answer was to fight one transgressive fantasy with another. Of course writing it was a risk; but then writing always is. Especially in this genre. If you write about fear, if you write about horror, if you write about hatred, pain, violence and death, the line between wallowing and making sense, between exploitation and exploration, between indulging or subverting, is narrow indeed. But what I do believe passionately is that over the last 25 years women have had to think harder about that line than men. And that in the process of thinking they have changed and challenged this rich, strange, complex, popular form of writing known as crime fiction, and that I am proud to count myself as one of them.

Notes

1 Raymond Chandler, *The Lady in the Lake* in *Later Novels and Other Writings* (New York: Library of America, 1995), pp. 38–9.
2 Philip Kerr, *A Philosophical Investigation* (London: Chatto & Windus, 1992), pp. 3–4.
3 Judith Walkowitz, *City of Dreadful Delight: Narratives of Sexual Danger in Late-Victorian England* (London: Virago, 1992).
4 *Birthmarks* (London: Michael Joseph, 1991); *Fatlands* (London: Hamish Hamilton, 1993); *Under my Skin* (London: Hamish Hamilton, 1995).

3

Fascination and Nausea: Finding Out the Hard-Boiled Way

David Trotter

In crime fiction, Auden said, the corpse must shock 'not only because it is a corpse but because, even for a corpse, it is shockingly out of place, as when a dog makes a mess on a drawing-room carpet.'[1] The really bad thing about murder, from one point of view, is that it makes a mess in a clean place. And yet that messiness, in Auden's view so crucial to stories about murder, so productive, rarely features in the explanations put forward for the broad and enduring appeal of crime fiction. Why?

The general view would seem to be that the shock detective fiction delivers is strictly hermeneutic. What we confront when the detective arrives at the scene of the crime is not a corpse, not a ruined carpet, but a ripening enigma. The enigma provokes interpretation rather than nausea: as the pursuit of meaning heats up, any shock we might initially have felt soon evaporates. Lodged securely in the empire of signs, tracking an infallible exegete through mean streets and even meaner discourses, we await closure. Most commentators feel that the detective story has proved resoundingly successful as a genre because it does what all stories do, only better: it masterfully produces, and at the same times trains us to produce, by its exuberant display of hermeneutic attitude, the 'fictive concords' which systematize existence.[2] Crime fiction has come to be regarded as 'the epistemological genre *par excellence*':[3] as a paradigm (an allegory) of narrative itself – or of the disciplinary techniques of surveillance and classification which, some would say, characterise the world we live in. In none of these accounts of the genre's weighty responsibilities does the corpse feature *as a corpse*, as a mess in a clean place.

One might think that there would be a difference, where corpses are concerned, between 'classic' and 'hard-boiled' detective fiction. Holmes, after all, has little affinity with decomposition, with waste matter. Were there to be a Baker Street dog, Mrs Hudson would surely dispose of its messes without interrupting his cocaine-daze. The hermeneutic foreplay which precedes each of his adventures is an announcement that, in Baker Street, sublimation is always already in progress. It textualizes matter (an abrasion, the scrape of mud on a boot). The corpses, too, are always already textual. They have been edited by assassination. Uncommon reader that Holmes is, a resident 'interpretive community' of one, or grudgingly two, they do not remain opaque (mere matter) for long. And they do not bring out an equivalent opacity in the interpreter; he knows doubt, on occasion, but never nausea, never shock. The fascination which stirs in him is chastely epistemological.

Philip Marlowe, by contrast, *does* have a certain affinity with waste matter. Murder might very well happen to him, and does happen repeatedly to his friends and associates. He might very well become redundant, he would have us believe, at any time, in one way or another. The best he can hope for, even on the least taxing of assignments, is to get messed up. Bent cops, old buddies, *femmes fatales*, barmen: they all have a shot at it. One way or another, in story after story, he will end up feeling that he has been treated like shit. Marlowe's problem is that he cannot distinguish himself sufficiently from the mess on the drawing-room carpet. And yet the commentaries insist that, in the 'hard-boiled' as in the 'classic' version of the genre, the detective's task is not crime prevention, but surveillance and control. The detective, it is said, 'fulfils the demands of the function of knowledge rather than that of lived experience: through him we are able to see, to know, the society as a whole, but he does not really stand for any genuine close-up experience of it.'[4]

The emergence in post-war American writing of a 'metaphysical' or 'anti-detective' detective fiction has reinforced the tendency to view the genre it imitates and burlesques in abstract terms.[5] 'Stories of detection and quest, emblems of the search-and-seizure mentality of the revelational plot (or, more aptly, plotted revelation), have been redesigned in contemporary fiction to conform to a modesty of aims enforced by an irremediably cryptic world.'[6] When those stories re-emerge in and through their deconstruction in books like Pynchon's *The Crying of Lot 49* (1966) and Auster's *The New York Trilogy* (1985–86), and in commentaries on Pynchon and Auster, they do so precisely as *emblems*. Reading them again, with Oedipa Maas or Daniel Quinn in mind, we recognize in them little except the 'search-and-seizure mentality' which supposedly informs them. Thus there is an entire tradition of commentary on

The New York Trilogy which represents detective fiction as no more than an embodiment of the literary and ideological closure Auster so elegantly refuses.[7] The criticism, unlike the book itself, it should be said, ignores mess altogether. Mess cannot be deconstructed (that is its great virtue).

This essay is dedicated to the proposition that Auden got it right: the corpse must shock, the reader (even, for a moment, the detective) must be shocked. My aim is not to produce a survey or a re-evaluation of the genre. It is, rather, to read a handful of 'hard-boiled' detective stories against the grain of a criticism which can no longer see a representation of material event for the discursive structure (the 'mentality') informing it.[8]

Floaters

After a few genteel preliminaries, Dorothy Sayers's *Whose Body?* (1923) unveils a body in a bath, in the suburban apartment belonging to an architect, Mr Thipps (henceforth 'poor little Thipps').

The body which lay in the bath was that of a tall, stout man of about fifty. The hair, which was thick and black and naturally curly, had been cut and parted by a master hand, and exuded a faint violet perfume, perfectly recognisable in the close air of the bathroom. The features were thick, fleshy and strongly marked, with prominent dark eyes, and a long nose curving down to a heavy chin. The clean-shaven lips were full and sensual, and the dropped jaw showed teeth stained with tobacco.[9]

This, seen from Wimsey's point of view, but without as yet any comment from him, is the closest the corpse comes to materiality. Its metamorphosis from material trace into sign-system begins almost immediately, as Wimsey deploys his monocle 'with the air of the late Joseph Chamberlain approving a rare orchid' (p. 18), and then accelerates, in chapter 2, as he reviews the evidence in the company of Parker of Scotland Yard. Wimsey has seen something Parker has not: a further, recessed materiality – wax in the ears, dried soap in the mouth, filthy toenails (pp. 33–4). The point of the embedded dirt, like that of the violet perfume and the long nose, is to emphasize by its opacity the speed and sureness with which Wimsey converts matter into sign. The clues are of two kinds: marks of sameness (race and class affiliation), which help to identify the victim; and marks of difference, which, by revealing what has been done to the victim's body (the way in which it has been edited), help to identify the killer. The dead body in the bath has become a set of signifiers, and what is signified, ultimately, is the identity of the killer. Chapter 2 ends with Wimsey's delighted recognition that 'we're up against a criminal – *the* criminal – the real artist and blighter with imagination' (pp. 34–5). He has looked straight through the victim (a blighter with no imagination, presumably) to an outline of the criminal (*the* criminal).

As Žižek points out, the scene which confronts the detective, like that which confronts the psychoanalyst, is often a 'false image' put together by the criminal in order to conceal or distort the evidence of his or her involvement in the crime. 'The scene's organic, natural quality is a lure, and the detective's task is to de-nature it by first discovering the inconspicuous details that stick out, that do not fit into the frame of the surface image.' The detective's and the analyst's domain, Žižek adds, is the domain of *meaning*, not of 'facts'. The crime scene is 'by definition "structured like a language"'. And since the identity of a signifier depends on its difference from other signifiers, the absence of a trait can have a positive value. 'Which is why the detective's artifice lies not simply in his capacity to grasp the possible meaning of "insignificant details", but perhaps even more in his capacity to apprehend absence itself (the nonoccurrence of some detail) as meaningful.'[10] Grasping the scene of the crime as a text edited by the murderer, the detective is able to apprehend him in and through his own deceptions, his own cunning, his fatal fondness for sly footnotes.

This is well said. But it fails to take account of the almost inevitable presence at the scene of someone other than the detective, someone who does not have the right hermeneutic attitude. In the interval between Wimsey's first glimpse of the body in the bath and his application of the monocle, poor little Thipps leaves the room. '"If you'll excuse me," he murmured,"it makes me feel quite faint, it reely does"' (p. 18). Poor little Thipps is about to barf. His nausea brings into play, at the very moment of detection, a curiosity about matter which Wimsey's brilliance soon overshadows, but never quite eclipses. If the monocle-flourishing detective authorises the pleasure and comfort we take in 'plotted revelation', then the queasy bystander reminds us of our vulnerability to shock.

In 'hard-boiled' fiction, Žižek adds, the detective loses the distance that would enable him 'to analyse the false scene and to dispel its charm'.[11] The detective's affinity with mess prevents him from converting matter into sign. To be sure, he is not likely to heave. But the scene charms him. It could conceivably charm him to the point of revulsion; its 'organic' quality, so hard to de-nature, could conceivably produce in him an 'organic' response. Acknowledging the detective's affinity with mess, 'hard-boiled' stories try harder than their 'classic' counterparts to make an inaugural mess. One indication of this concerted effort is their regular deployment of floaters. Floaters, the result of death by drowning, disrupt the conversion of matter into sign. They are, as everyone must know by now, hard to fingerprint and thus to identify. They evade the very technique which first associated detection (and detective stories) with the romance of scientific reason, of power/knowledge. The lady Philip Marlowe finds in a lake turns out not to be the lady he thought she was.

> The depths cleared again. Something moved in them that was not a board. It rose slowly, with an infinitely careless languor, a long dark twisted something that rolled lazily in the water as it rose. It broke surface casually, lightly, without haste.

The floater is a slow intrusion from a world beyond this one: it has to be prised out of or detached from an element which is not the earth we tread or the air we breathe. It is an immanent shapelessness.

> The thing rolled over once more and an arm flapped up barely above the skin of the water and the arm ended in a bloated hand that was the hand of a freak. Then the face came. A swollen pulpy gray white mass without features, without eyes, without mouth. A blotch of gray dough, a nightmare with human hair on it.[12]

Marked off within the text as a separate paragraph, falling below the threshold at which devices like punctuation can be expected to work properly ('A swollen pulpy gray white mass ...'), this account reaches a discursive limit. The scene into which the floater emerges is very definitely not 'structured like a language'. The distinctions which found the symbolic (inside/outside, nightmare/reality) no longer obtain. The water may possess a 'skin', but the corpse which rolls out of it does not. Once again, a hermeneutically challenged bystander offers a point of view other than that of the detective. 'The man called Andy got a dusty brown blanket out of the car and threw it over the body. Then without a word he went and vomited under a pine-tree.' (p. 53) In this case, furthermore, Marlowe himself proves vulnerable, albeit belatedly. He has a nightmare about a blotchy, water-logged corpse, about putrescence (p. 90).

Floaters are perhaps too much of a gift to writers trying to hard-boil discursive structure out of the crime scene. Marlowe's nightmare with human hair on it re-surfaces in Joseph Koenig's *Floater* (1986), where a Florida fisherman gets rather more than he had bargained for.

> Forgetting about alligators, he stepped onto some flat rocks and snagged the floating object. It was the body of a woman – about fifty, he guessed. He didn't care to look at it closely. Something had nibbled at the arms and legs, and the torso was bloated and distended. He tried to nudge it to shore but barely moved it. Then he noticed a dark mass clinging to what was left of an arm just below the surface. He poked at it with the rod, and the body submerged and bobbed ponderously.[13]

This corpse, too, gets a paragraph to itself. But the fussy notation of separate movements (stepping onto the flat rocks, nudging, poking) ensures that the floater will not float in anyone's dreams. The fisherman has been endowed with the detective's more-or-less stalwart indifference to organicism, but not his interpretative zeal. Nobody barfs.

Altogether more effective are the floaters whose retrieval punctuates Thomas Harris's *The Silence of the Lambs* (1988). Clarice Starling's first assignment is to print one, in a funeral parlour in West Virginia. Nobody barfs: but only because kindly old Jack Crawford, head of the Behavioural Science section at the FBI,[14] has brought his 'Vicks VapoRub' with him. This corpse yields a rich crop of clues: the pattern of the flaying, the position of the entry wound, the broken fingernails, and, above all, the death's head moth lodged in the throat. But its metamorphosis from material trace into sign-system remains incomplete. Just as Marlowe has nightmares about the lady in the lake, so the memory of the funeral parlour comes back to Starling, in tension with the 'savage pleasure' of problem-solving (p. 307), at once overwhelming her and re-kindling her rage. The corpse endures, not only as a set of signifiers, a spectacle deceptively staged by an assassin, but as affect, as shock. This floater floats. It becomes a rhythm, a refrain: '... when Catherine Martin floats, when the next one floats, and the next one floats...' (p. 175). At such moments the savage pleasure of problem solving collapses in on itself.

Floaters are the genre's admission that matter shocks and that shocks matter. Even when the corpse does not float, it must make a mess, and people must respond to that mess as they usually respond to messes, with disgust. Auden got it right, and Hollywood has acknowledged that he did by incorporating what one might call the barf-scene into the conventions of *film noir.* There is a memorable one in Richard

Marquand's *Jagged Edge* (1985), for example, when the District Attorney first enters the bedroom where a publishing magnate's wife lies battered to death. By now, the convention is so well established that the Coen brothers' *Fargo* (1996) even includes a meta-barf-scene. Police chief Marge Gunderson, inspecting the overturned car which contains one of the victims in a multiple homicide, announces that she is going to barf; but what turns her stomach is morning sickness, not the sight of blood.

The corpse is detective fiction's first and only object, its defining obstacle: the veil of matter whose piercing, with the help, it may be, of archives and laboratories, announces that gnosis has begun. Thereafter, out of earshot of commonplace retching, of the reader's retching, the 'cognitive hero'[15] enters a virtual reality where even the beatings and the seductions are patterns thrown on a screen, events staged by the first and only assassin (the implacable Other). Is she, or he, subject enough to survive? Fictional detectives have to be more characterful than fictional spies or fictional cowboys: interiority is expected of them. They go to work with their subjecthood, and *on* their subjecthood, for they must not be deflected from the hermeneutic endeavours upon which, in a fallen world, truth and justice depend. The sublimation which converts a dead body into a set of clues significantly prefigures the detective's long and arduous sublimation of his or her own body: of desire and rage, of fear and loathing. Here, too, however, the sublimation may not be complete. Clarice Starling answers all of Dr Hannibal Lecter's probing questions, except the last one. 'The other thing I wonder is ... how do you manage your rage?' (p. 163). It is a question to which the *genre* has no answer. I want to define that inability by examining the detective's relation to authority, and to desire.

Fathers

The Silence of the Lambs and *The Big Sleep* are coming-of-age sagas in which the protagonist has to win approval in the eyes of the father (society), without compromising his or her independence. But which father? In each case, there is a good father, wise and incorruptible; and a bad father, wise and corrupt. The good father has the power to grant or withhold a recognition which is not his alone, and which does not issue directly from him. He is childless (he has not yet appointed his successor). His wish is for the survival of the race, of the social order. The bad father grants or withholds personal recognition: whatever he gives issues directly from him, in the form of wealth or information.

He has been recklessly fertile: his misbegotten children are always already wreaking havoc. His wish is to populate the universe, for the duration of his own life, with people who interest him: he knows, and gives shape to, neediness (desire and rage, fear and loathing).

In *The Big Sleep*, the good father is genteel, solitary Taggart Wilde, the District Attorney, for whom Marlowe once worked; the bad father is General Sternwood, begetter of two uncontrollable daughters, and source of dubious largesse. In *The Silence of the Lambs*, the good father is genteel, solitary Jack Crawford, Head of Behavioural Science; the bad father is Dr Hannibal Lecter, confessor to serial killers, and source of vital information. Wilde and Crawford see a place for their proteges in the system they uphold. Sternwood and Lecter identify and exploit neediness. The power of these stories is that they divert the rite of passage into and through a relation with the bad father in which fascination grips with ever increasing intensity until it finally reaches and exceeds its limit. Both open with a visit to the bad father.

In chapter 1 of *The Big Sleep*, Marlowe arrives at General Sternwood's Bel-Air mansion. 'I was everything the well-dressed private detective ought to be. I was calling on four million dollars.' The narrative enacts with almost pedantic thoroughness the intensity of his engrossment by wealth, and by the corruption wealth entails. He passes through the grandiose hall, where Carmen Sternwood flirts with him and falls against him, round the lawn behind it, to the greenhouse where the General sits, in absolute recession, at the heart of darkness. The greenhouse is the mansion's primary cell, a 'deep' space adapted for its owner's exclusive use, a shrine; to reach it from the street, Marlowe must negotiate a whole series of 'shallow' intermediary zones.[16] And yet the shrine's occupant does not exactly radiate power. 'Here, in a space of hexagonal flags, an old red Turkish rug was laid down and on the rug was a wheel chair, and in the wheel chair an old and obviously dying man watched us come.' (p. 13) The syntax mimics a roundabout and carefully policed approach to the source; but the source is dry. At once shrunken and diffuse, impotent and promiscuous, General Sternwood exercises power by metastasis. It is too late, we soon gather, for cognitive heroism. What mediates Marlowe's encounter with corrupt authority is not the romance of reason, but fascination and nausea. The orchids surrounding the wheelchair, through which he must force a path, make him feel sick. 'The plants filled the place, a forest of them, with nasty meaty leaves and stalks like the newly washed fingers of dead men.' (p. 13) Here, at this proleptic crime scene (or morgue scene), the cognitive hero almost barfs. Indeed, cognition is not really the issue.

There is nothing to interpret. Marlowe goes to the bad father, a shameless hedonist, to receive acknowledgement of his own desire for wealth and power; but the bad father has already given all he has to give, or almost all. Those nasty meaty leaves mark fascination's limit: the point at which it becomes aware of itself, for the first time, in its own excess.[17]

In chapter 2 of *The Silence of the Lambs*, Clarice Starling arrives at the Baltimore State Hospital for the Criminally Insane to interview Hannibal Lecter. Lecter's cell is the most recessed room in the most recessed part of a recessed institution. Like General Sternwood, he has been immobilised (though not for very long). To reach this primary space, and have her own desires and hatreds acknowledged by the shamelessly desiring and hating bad father, Starling, like Marlowe, must placate the functionaries (Dr Chilton, of the 'fast grabby eyes' (p. 11)), and pass through the intermediate zones (pp. 11–12). Marlowe has to cope with orchids, Starling with Miggs, who flicks semen at her as she returns from Lecter's cell (p. 23). Again, there is not much scope for cognitive heroism. As Lecter recognizes, cognition, the ostensible purpose of Starling's visit, is not really the issue. Like the orchids, Miggs's mess marks a limit: fascination's limit in, or at, nausea.

If these stories begin with the bad father, in deep space, already at the limit, they conclude with the dispersed violence of his errant children. Here, too, the vocabulary of revulsion intrudes where one might not necessarily have expected it. On his first visit to the Sternwood mansion, Marlowe looks out over the oilfields which are the source of the family fortune. 'The Sternwoods, having moved up the hill, could no longer smell the stale sump water or the oil, but they could still look out of their front windows and see what had made them rich.' (pp. 25–6) At the end of the story, the saving distance articulated by sight collapses. Marlowe goes down among the stinking oilfields to confront, in one of Chandler's most Conradian moments, the General's wild child, Carmen.

> The wells were no longer pumping. There was a pile of rusted pipe, a loading platform that sagged at one end, half a dozen empty oil drums lying in a ragged pile. There was the stagnant, oil-scummed water of an old sump iridescent in the sunlight.
>
> 'Are they going to make a park of all this?' I asked.
>
> She dipped her chin down and gleamed at me.
>
> 'It's about time. The smell of that sump would poison a herd of goats. This the place you had in mind?'
>
> 'Uh-huh. Like it?' (p. 209)

The sump, we learn later, contains the body of Rusty Regan, the General's other 'son': by now, presumably, a 'horrible decayed thing', as Vivian Sternwood will put it (p. 218). That the sump conceals what Marlowe has been looking for from the start matters less, at this point, than that it should nauseate him. There will be time enough, after nausea, to find out who ended up where. For the encounter with the bad father and his works is not about truth and justice; it is about fascination, and fascination's excess.

Clarice Starling must confront Hannibal Lecter's most malevolent 'son', Jame Gumb, in his burrow: a maze of underground rooms containing a variety of horrible decayed things.

> In absolute black the hiss of steam pipes, trickle of water.
> Heavy in her nostrils the smell of the goat. (p. 332)

The stench reminds Starling, as it reminds Marlowe, that she has looked too hard for too long, that at some time or another, in some way, she has colluded with the bad father who knows exactly what she needs. As with the floater in *The Lady in the Lake*, an interruption to the narrative, a surplus of rhetorical effect over syntax, administers shock to the reader (somewhat portentously, in this case).

In *The Big Sleep* and *The Silence of the Lambs*, the good father supervises the hermeneutic process which is generally taken to characterise the genre of detective fiction. Marlowe does his detecting for Wilde's benefit (pp. 108–9), Starling for Crawford's (pp. 151–2). With the bad fathers, another scheme comes into play, one the genre is not thought to favour. In this case, complicity establishes a dialectic of fascination and nausea which forms (or deforms) the protagonist more comprehensively than involvement in the hermeneutic process, or in disciplinary techniques of surveillance and control. The dialectic, we might note, outlasts the tale told; it prevents closure. The bad father's even worse children die, or are put away; the bad father survives.

Quasi-celibacy

The dialectic activated by Marlowe's encounter with General Sternwood takes shape in his subsequent encounters with Vivian and Carmen, who together constitute a split or doubled *femme fatale*. Žižek points out that the detective's fate (his acknowledgement of the scope and intensity of his own desires) is usually decided at the moment when the *femme fatale* suffers her final breakdown. This moment, when her power of fascination suddenly 'evaporates', leaving him (and us) with little but disgust, confirms either his triumph or his collapse. What awaits her beyond hysteria is the death drive at its purest. 'When the woman reaches this point, there are only two attitudes left to the man: either he "cedes his desire", rejects her, and regains his imaginary, narcissistic identity (Sam Spade at the end of *The Maltese Falcon*), or he *identifies* with the woman as symptom and meets his fate in a suicidal gesture (the act of Robert Mitchum in what is perhaps the crucial *film noir*, Jacques Tourneur's *Out of the Past*).'[18] In *The Big Sleep*, Chandler separates the fascination from the disgust. From the moment Vivian Sternwood first arranges her legs on the *chaise-longue*, she is 'worth a stare' (p. 22). Carmen, on the other hand, occasions neither 'embarrassment' nor 'ruttishness', even when stark naked (p. 40). Her collapse, when it comes, renders her utterly abject: she froths at the mouth, and wets herself (p. 211). The intensity of the disgust Marlowe feels for Carmen enables him to preserve his narcissism intact, while still remaining vulnerable, it may be, or might be under somewhat different circumstances, to a stare-worthy woman like Vivian. Marlowe's erotic status, like that of many other private eyes, male and female, is one of quasi-celibacy. These men and women are sexually suspended: to vivid effect.

Reasons have been advanced for attributing Marlowe's suspension to homosexuality.[19] In chapter 24 of *The Big Sleep*, Marlowe finds Carmen Sternwood in his bed, naked. 'It's so hard for women – even nice women – to realize that their bodies are not irresistible.' (p. 153) When he declines the offer, she hisses at him, 'her mouth open a little, her face like scraped bone' (p. 153), and calls him a 'filthy name' (p. 154). 'I couldn't stand her in that room any longer. What she called me only reminded me of that.' (p. 154) Later, after she has gone, seeing the imprint of her 'small corrupt body' on the sheets, he tears the bed to pieces savagely (p. 155). The next morning, he wakes up with a 'hangover' from women. 'Women made me sick.' (p. 156) The sickness at women may perhaps indicate, it could be said, not a generic requirement, but a 'homosexuality' specific to Marlowe.

I would argue, on the contrary, that Marlowe's revulsion from Carmen is a development of the dialectic of fascination and nausea established by his encounter with the bad father. Marlowe is fascinated by the bad father's corrupt wealth, and by the corruption, sexual and otherwise, of his two daughters. Vivian and Carmen Sternwood are figures of transgression, of a non-pathological desire which has assumed responsibility for its own fate. The novel's evident fascination with homosexuality is a fascination with a sexuality identified, by the culture which disowns it, and by Chandler himself, as transgressive. The 'horror' some people feel at homosexuality, Chandler wrote, is like a woman's fear of scorpions. The bonds which 'hold us to sanity' are fragile, and constantly under threat from 'repulsive insects' and 'repulsive vices' alike. 'And the vices are repulsive, not in themselves, but because of their effect on us. They threaten us because our own normal vices fill us at times with the same sort of repulsion.'[20] If, as seems likely, the 'filthy name' Carmen calls Marlowe is 'faggot', then she thereby associates herself with the repulsive insects and repulsive vices which threaten his hold on sanity (on narcissism). Carmen herself, with her hissing and her scraped-bone look, resembles a snake rather than a scorpion. Her bestiality and the filthy name she calls Marlowe constitute her as the point at which fascination (those 'normal vices', perhaps) becomes aware of itself in its own excess. Vivian Sternwood describes Proust to Marlowe as a 'connoisseur in degenerates' (p. 58). So was Chandler. Chandler used the idea of 'the homosexual' to define degeneracy: that excess of non-pathological desire which must be pathologized, through the nausea it produces, if 'normal vices' are to remain normal. The genre demanded it of him.

Marlowe's curious quasi-celibacy is very much at issue in the scene in the Fulwider Building, when he overhears Lash Canino, Eddie Mars's hit man, killing honourable, diminutive Harry Jones. The building itself is as 'nasty' (p. 166) as General Sternwood's orchids.

> The fire stairs hadn't been swept in a month. Bums had slept on them, eaten on them, left crusts and fragments of greasy newspaper, matches, a gutted imitation-leather pocketbook. In a shadowy angle against the scribbled wall a pouched ring of pale rubber had fallen and had not been disturbed. A very nice building. (p. 167)

Marlowe does not frequent mean streets and nasty lobbies in order to feed a 'habit of loathing'.[21] He frequents them because he believes he will find there an acknowledgement of his own desires. What he finds instead is waste matter, a mess which could quite easily be a mess he himself has already made (the ring of pale rubber already 'pouched'). He finds a pathologized version of his own desires. And yet he is not quite ready to quit. In this 'very nice' building, something 'very nice' happens. The delicious guilt Marlowe feels as a man dies on the other side of a flimsy partition is as intense a pleasure as any he experiences in the novel. William Marling has compared the Fulwider Building to one of Deleuze and Guattari's 'celibate machines': a mechanization of experience which produces unpredictable and uncontrollable autoerotic intensities compounded of pleasure and pain.[22] Marlowe subsequently discharges the intensity stored autoerotically during Jones's murder when he shoots Canino. 'But his gun was still up and I couldn't wait any longer. Not long enough to be a gentleman of the old school. I shot him four times, the Colt straining against my ribs.' (p. 194)

Jones's murder strikes an odd note. Canino carries a gun, but instead of pulling the trigger, like any other Chandler hoodlum, he persuades Jones to drink from a bottle of poisoned whisky which he just happens to have with him. It is as though he seduces Jones to death. 'The purring voice was now as false as an usherette's eyelashes and as slippery as a watermelon seed' (p. 170). The connection I would make, taking a hint from Vivian Sternwood's reference to Proust, is with the scene in *A la recherche du temps perdu* when the narrator overhears the Baron de Charlus seducing Jupien. He cannot believe how noisy they are: it is almost as though a murderer and his resuscitated victim were taking a bath together, in order to wash away the traces of the crime. He concludes that pleasure is the one thing in life as vociferous (as intense) as pain, especially when it involves an immediate concern for cleanliness.[23] Proust's narrator overhears a seduction which sounds like a murder; Marlowe overhears a murder which sounds like a seduction. The implication of both scenes is that the intensity of a transgressive experience can only be measured by the mess it makes. Entering the room where Harry Jones lies, after Canino's departure, Marlowe sees that he has vomited on his overcoat (p. 171). Like the narrator in *A la recherche*, the hard-boiled detective is there, on our behalf, to experience rather than to know: to measure out, from a point inside it, the dialectic of fascination and nausea.

Notes

1 'The Guilty Vicarage', in *The Dyer's Hand* (London: Faber and Faber, 1948), p. 151.

2 The phrase 'fictive concords' is Frank Kermode's: *The Sense of an Ending: Studies in the Theory of Fiction* (London: Oxford University Press, 1967), pp. 7 and 18.

3 Brian McHale, *Constructing Postmodernism* (London: Routledge, 1992), p. 147. According to Peter Huehn, the genre 'thematizes narrativity itself as a problem, a procedure, and an achievement': 'The Detective as Reader: Narrativity and Reading Concepts in Detective Fiction', *Modern Fiction Studies*, 33 (1987), 451–66, here p. 451.

4 Frederic Jameson, 'On Raymond Chandler', in J.K. van Dover (ed.), *The Critical Response to Raymond Chandler* (Westport, Conn.: Greenwood Press, 1995), pp. 65–87, here pp. 69–70.
5 Michael Holquist, 'Whodunit and Other Questions: Metaphysical Detective Stories in Post-War Fiction', *New Literary History*, 3 (1971), 135–56; William V. Spanos, 'The Detective and the Boundary: Some Notes on the Postmodern Literary Imagination', *Boundary*, 2.1 (1972), 147–68.
6 Arthur M. Saltzman, 'De(in)forming the Plot', in *Designs of Darkness in Contemporary American Fiction* (Philadelphia: University of Pennsylvania Press, 1990), pp. 52–95, here p. 52.
7 Saltzman, 'De(in)forming the Plot', pp. 56–70; Alison Russell, 'Deconstructing *The New York Trilogy*: Paul Auster's Anti-Detective Fiction', *Critique*, 31 (1900), 71–84; Norma Rowan, 'The Detective in Search of the Lost Tongue of Adam: Paul Auster's *City of Glass*', *Critique*, 32 (1991), 224–34; William Lavender, 'The Novel of Critical Engagement: Paul Auster's *City of Glass*', *Contemporary Literature*, 34 (1993), 219–39; Steven E. Alford, 'Mirrors of Madness: Paul Auster's *The New York Trilogy*', *Critique*, 37 (1995), 17–33, and 'Spaced Out: Signification and Space in Paul Auster's *The New York Trilogy*', *Contemporary Literature*, 36 (1995), 613–32; Carl D. Malmgren, 'Detecting/Writing the Real: Paul Auster's *City of Glass*', in Theo D'haen and Hans Bertens (eds), *Narrative Turns and Minor Genres in Postmodernism* (Amsterdam: Rodopi, 1995), pp. 177–201; Madeleine Sorapure, 'The Detective and the Author: *City of Glass*', in Dennis Barone (ed.), *Beyond the Red Notebook: Essays on Paul Auster* (Philadelphia: University of Pennsylvania Press, 1995), pp. 71–87; Jeffrey T. Nealon, 'Work of the Detective, Work of the Writer: Paul Auster's *City of Glass*', *Modern Fiction Studies*, 42 (1996), 91–110.
8 Slavoj Žižek attempts something similar, to brilliant effect, in 'Two Ways to Avoid the Real of Desire', in *Looking Awry: An Introduction to Jacques Lacan through Popular Culture* (Cambridge, Mass.: MIT Press, 1991), pp. 48–66. I discuss the bodies in 'classic' detective fiction in *The English Novel in History 1895–1920* (London: Routledge, 1993), pp. 220–7.
9 Dorothy L. Sayers, *Whose Body?* (London: Coronet Books, 1989), pp. 17–18.
10 Žižek, 'Two Ways', pp. 53 and 57.
11 Ibid., p. 60.
12 Raymond Chandler, *The Lady in the Lake* in *Later Novels and Other Writings* (New York: Library of America, 1995), pp. 38–9.
13 Joseph Koenig, *Floater* (New York: The Mysterious Press, 1986), p. 153.
14 Thomas Harris, *The Silence of the Lambs* (London: Mandarin, 1990), p. 79.
15 Wladimir Krysinski, *Carrefours de signes: Essais sur le roman moderne* (The Hague: Mouton, 1981), p. 168.
16 Raymond Chandler, *The Big Sleep* (Harmondsworth: Penguin Books, 1978), p. 9. I derive this understanding of the social relations embodied in the structure of a building, very crudely, from Bill Hillier and Juliette Hanson, *The Social Logic of Space* (Cambridge: Cambridge University Press, 1984), a study whose rigour and sophistication vastly exceeds the scope of my own argument about General Sternwood's greenhouse.
17 I am indebted to Maurice Blanchot's account of fascination as the opening to an 'outside' which 'has no location and affords no rest', in *The Space of Literature*, trans. Ann Smock (Lincoln: University of Nebraska Press, 1982),

pp. 30–3. For Blanchot, corpses belong definitively to that outside. 'The corpse is a reflection becoming master of the life it reflects – absorbing it, identifying substantively with it by moving it from its use value and from its truth value to something incredible – something neutral which there is no getting used to.' (p. 258) Corpses also feature in Julia Kristeva's theory of abjection, a theory whose relevance to detective fiction can only have been obscured by the prevailing concern with epistemology: *Powers of Horror: An Essay on Abjection*, trans. Leon S. Roudiez (New York: Columbia University Press, 1982), pp. 3–4.

18 Žižek, 'Two Ways', pp. 65–6.

19 Michael Mason, 'Marlowe, Men and Women', in Miriam Gross (ed.), *The World of Raymond Chandler* (London: Weidenfeld and Nicolson, 1977), pp. 90–101.

20 Letter of 9 July 1949 to Dale Warren: Raymond Chandler, *Selected Letters* (ed.) Frank MacShane (London: Jonathan Cape, 1981), p. 185.

21 Tom S. Reck, 'Raymond Chandler's Los Angeles', in *The Critical Response*, pp. 109–15, here p. 111.

22 William Marling, *The American roman noir: Hammett, Cain, and Chandler* (Athens, Georgia: University of Georgia Press, 1995), pp. 213–16.

23 Marcel Proust, *A la recherche du temps perdu*, translated by C.K. Scott Moncrieff and Terence Kilmartin as *Remembrance of Things Past*, 3 vols (Harmondsworth: Penguin, 1983), Vol. II, p. 631.

4

The Writers Who Knew Too Much: Populism and Paradox in Detective Fiction's Golden Age

David Glover

'I do wish to Heaven you had given us more of these books,' wrote Dorothy L. Sayers to fellow crime writer E.C. Bentley in April 1936. The occasion was the appearance of the long-awaited sequel to *Trent's Last Case*, Bentley's enormously influential detective novel published in 1913. Sayers' gushing enthusiasm – 'yours are BOOKS, full of humanity and the Humanities, touching life on all sides' – is tempered with a certain regret, a plaintive note of reproach at Bentley's prolonged silence. 'With you to help us', Sayers chides, 'we should not have taken half so long to get the detective novel recognized as literature.'[1] Bentley had written too little, too late.

Yet the dates, 1913–1936, not to mention the triumphant sense of achievement, are significant. In Sayers' celebratory tone one begins to hear the first stirrings of a now familiar theme: that these years represent an unparalleled 'Golden Age' in the annals of detective fiction, far outstripping even the formative work of Conan Doyle or Poe. According to this view, given canonical status in Howard Haycraft's classic survey *Murder for Pleasure: The Life and Times of the Detective Story* (1941), the 1920s and the 1930s were a period of growing maturity, an era of consolidation and technical mastery, for which Bentley, in *Trent's Last Case*, had deftly provided the stylistic inspiration and literary model. In its most euphoric version, the Golden Age was extolled as 'the richest single age in the literature', reaching new heights of ingenuity and sophistication and winning new readers to the fold.[2] After 1936 one might be forgiven for thinking that the future of the genre was secure.

Of course, in practice the development of detective fiction was never so straightforwardly linear nor so self-directed as this simplified picture of the Golden Age suggests, and nor were its authors quite as united as

Sayers's all-inclusive 'we' rhetorically implies. For despite efforts to police the genre by stipulating the 'very definite laws' upon which 'the writing of detective stories' might be said to depend, crime fiction has always been parasitic upon a host of other investigative narratives, controversial epistemologies of the social body ranging from forensic medicine to criminology and political science, with their unruly and disquieting objects of inquiry.[3] Given the much disputed status and ramifications of many of these discourses, this has meant that writers have often differed profoundly in how they have conceived the field in which they were working, where they have positioned themselves within it and what they thought they were doing there. In this essay I want to try to recapture the stubborn cross-purposes, the sharp divergences of emphasis and aspiration among crime writers that the ideology of the Golden Age was, in part, designed to smooth over or suppress in order both to defend the genre against condescension and dismissal and to raise its standing in the wider world of letters.

What made the detective fiction written in the inter-war years seem like an advance upon that of previous generations? Looking back over the history of the genre, commentators have typically given pride of place to two main characteristics. The first, briefly mentioned already, is that detective fiction comes to be codified in terms of a more or less agreed set of rules which, whether honoured or not, set up an expectation that readers will have a real possibility of solving the mystery that the author has devised. And second, the novel definitively replaces the short story as the literary space in which this contest between author and reader takes place – hence the importance of *Trent's Last Case* as the prototype of a more sustained, more accomplished kind of writing. Put like this, it might sound as if the Golden Age was merely a new order of gamesmanship – detective fiction as fun, albeit a more elaborate type of diversion than hitherto. Yet while this is certainly one very widespread way of viewing the Golden Age, it also begs the question of just how serious this fun was meant to be. After all, playing the game has always been a relatively serious matter in English culture. The history of the Golden Age is, I want to suggest, a history of competing modes of seriousness, of disagreements about what is worth taking seriously, and why. There is, to take the most obvious example, a palpable tension between the oft-maligned ideal of the 'crime-puzzle, pure and simple' (Anthony Berkeley) and 'the easy touch of the real craftsman with centuries of civilization behind him' (Dorothy L. Sayers), an antinomy that echoes much older, but no less intractable, cultural oppositions such as the division between science and the humanities, or between reason and intuition.[4]

The project of legislating rules for the genre is a case in point. The most famous of these were published by practitioners themselves, notably S.S. Van Dine and Monsignor Ronald A. Knox, each of whom independently published articles on the subject in 1928; and a similar list of maxims formed part of the oath sworn by incoming members of the Detection Club, founded in London in the same year. But one of the earliest ventures of this kind had already appeared in print more than 12 months before. In a review essay published in *The New Criterion* of January 1927, T.S. Eliot argued that the 'rising demand' was 'producing a different, and as I think a superior type of detective story' and that it was now possible to enumerate 'some general rules of detective technique' by which this kind of fiction might be judged.[5] Today Eliot's 'rules of the game' sound familiar enough, but it is worth noting that they are far less playfully phrased than, say, Knox's 'Detective Story Decalogue' with its stipulation that 'No Chinaman must figure in the story'. ('Why this should be so I do not know,' Knox quickly and disarmingly adds.)[6] According to Eliot, detective fiction should eschew 'elaborate and incredible disguises', 'occult phenomena' or 'mysterious and preposterous discoveries made by lonely scientists' and should try to ensure that the skills of the detective and the motives of the criminal are within the grasp of ordinary human consciousness.[7] The pathological mind and the extremes of genius are equally to be avoided.

As one might expect, Eliot's criteria are fastidious and severe. They are designed not only to distinguish between good and bad, but also to expel false claimants from the review pages. By Eliot's own exacting standards, three of the nine books he considers 'are not properly detective stories' at all and every one of the remaining six 'violates' one rule or another. Perhaps this is only to be expected in a situation where 'the output of detective fiction has increased rapidly'; but it may also be read as an attempt to control the epistemological disorder that always threatens to radically disrupt the rational structures of the genre.[8] As Uri Eisenzweig has shrewdly observed, the constitutive features of the detective story – 'silences and testimony, masks and clues, mystery and truth' – 'inevitably open up the text to *other* discursive spaces.'[9]

As I noted above, Eliot insisted that in any detective fiction worth its salt, the 'character and motives of the criminal should be normal', while the detective must be 'highly intelligent but not superhuman.' To fail to adhere to these guidelines is to expose the detective story to 'an irrational element...which offends us.'[10] But what happens when such elements are deliberately intruded into the plot? To help think the unthinkable I want to turn to a particularly troubling text by

G.K. Chesterton, one of the best-known and highly regarded of detective fiction writers. In 1922 Chesterton produced a book that, apart from its title (which was stolen and popularized by Alfred Hitchcock), is almost forgotten today: *The Man Who Knew Too Much*. It consists of eight apparently unrelated mystery stories which gradually coalesce into a larger narrative and which builds to a devastating climax in the final tale. The book introduces a puzzlingly cynical and distinctly well-connected English gentleman called Horne Fisher, who appears to see through everyone and everything, but who, in a typically Chestertonian paradox, is unable to act upon what he knows. In short, the burden of his knowledge is precisely what qualifies him as 'the man who knew too much.' Now within the epistemology of the Golden Age it is *de rigueur* that nothing will be what it seems. Yet in these stories the suspicion of appearances extends all the way up the social ladder to the Prime Minister who, in one story, is revealed as the murderer. And, in one last devastating paradox, partially anticipating Agatha Christie's stratagem in *The Murder of Roger Ackroyd* (1926), the detective also numbers himself among the guilty men, for Horne Fisher confesses to having killed his own uncle. Perhaps worst of all, although every mystery in the book is cleared up, none of the criminals can be brought to justice.

Surely in this terminal *reductio ad absurdum* the detective story has definitively reached its limits? What possible structure of 'normal' motivation could satisfactorily explain such a morass of intrigue and conspiracy? Is this in fact a negative variant of the classic detective story in which crimes are covered up instead of the world being put to rights? In reality, the intimation of an endless series of cover-ups alerts us to the strong political overtones of the book that press in upon the mystery from the very beginning. On the first page of the first story we learn that Horne Fisher's Watson, his interlocutor Harold March, a 'rising reviewer and social critic', is a man badly in need of political re-education: he was 'the sort of man who knows everything about politics; and nothing about politicians.' Indeed, he knew a good deal 'about almost everything, ... except the world he was living in.'[11]

What pushes *The Man Who Knew Too Much* to the edge of the genre, then, is its contamination by a peculiarly sordid brand of politics, and over and over again the narrative fetches up against a paralyzing, over-informed disillusion. Fisher, smiling 'his dreary smile' (p. 27), actually tells March at one point that his understanding of realpolitik makes him prefer to 'moon away my time over things like stinking fish' (p. 31). But politics does not immediately displace detective technique;

and even at the final denouement 'the steps of deduction' are still there, though Fisher doubts whether March will 'want to listen to them' (p. 183). Nevertheless, in the closing story, 'The Vengeance of the Statue', not only is there a moment of retribution, but Horne Fisher finds himself compelled to *act*. As the book draws to an end the insidious political sub-text explodes into a major international conflict as England is invaded by an unnamed Eastern power. Summoned to this apocalyptic battle, Horne Fisher takes up arms and dies a hero's death to save his country, learning a different order of wisdom as he does so. Lying 'in the shadow at the foot of the ridge, as stiff as the stick of the fallen rocket', he is transformed: for once 'the man who knew too much knew what is worth knowing' (p. 190).

The motivation for the crimes that give the book its meaning are both bizarre and hideously 'normal': 'the seamy side of things; all the secret reasons and rotten motives and bribery and blackmail they call politics' (p. 76). England's politicians have sold out their country to the forces of alien capital, permitting the 'dexterous little cosmopolitan guttersnipe' to take on the disguise of 'a dashing gentleman and a worthy business man and a philanthropist and a saint' (p. 29). Fisher's patriotism is awakened when he recognizes that 'we have yielded to foreign financiers so long that now it is war or ruin.' The nation has been infiltrated by a system of 'infernal coolie capitalism', immigrant Chinese labour brought into the country 'with the deliberate intention of reducing workmen and peasants to starvation' (p. 172). Fisher had hoped to build a 'yeoman party' with the 'idea of championing a new peasantry against a new plutocracy' (p. 146), but it is too late for that. England has been mortgaged lock, stock and barrel to 'a gang of infernal Jews' and now 'all hell [is] beating up against us, simply because Nosey Zimmern has lent money to half the Cabinet' (p. 89). This is the most heinous crime imaginable, the guilty secret which induces all the other crimes in the book; and it is in these murky waters that the base criminal motives at work in *The Man Who Knew Too Much* are to be found. In an England that has been enslaved by an 'alien' or 'cosmopolitan' coterie whose true identities are barely concealed by anglicized names the only possible solution is blood sacrifice.

At this point one might suspect that Chesterton was rather desperately borrowing from the detective story's less respectable cousin, shamelessly rifling through the detritus of those political thrillers from the turn of the century whose stock in trade were tales of invasion and Jewish conspiracy. And there is much truth to this contention. John Buchan's *The Thirty-Nine Steps* (1915) is, for example, haunted by the possibility that 'the Jew' is 'the man who is ruling the world just now.'[12] More disturbingly, however, one could also read Chesterton's paranoid inflection of the genre as a demand that the mystery story *give way* to the political thriller, forcing the armchair detective to yield to the imperial martyr. Or, put another way, we could perhaps see this disruption of the protocols of detective fiction as a sign of the imagined depth of England's national crisis. Here antisemitism ceases to be a casual or incidental aspect of the detective story and generic boundaries begin to dissolve. 'Criminal crimes are soothing, adventure crimes are frightening', wrote Gertrude Stein in her attempt to explain 'Why I Like Detective Stories' (1937).[13] But when the adventures of a popular militia seem to offer the only solution to 'criminal crimes', as happens in *The Man Who Knew Too Much*, then what was once soothing can rapidly become utterly terrifying.

A Golden Age purist like Eliot might conceivably have argued that the political irrationalism of Chesterton's text was intimately linked to the author's willingness to break the rules: the introduction of oriental labourers into the story, in flagrant contravention of Knox's fifth rule banning Chinamen, was a sure sign of trouble. But the respectability of antisemitism in this period, not least in Eliot's own work, suggests that things were not so simple.[14] The *cordon sanitaire* that was supposed to separate 'detective technique' from the adventure tale was far from being politically innocent. To see why, let us turn to a very different case, that of R. Austin Freeman.

Eliot judged Freeman's novel *The D'Arblay Mystery* (1926) to be the best of the books under review, calling it 'the most perfect in form'.[15] Whether or not this is strictly accurate – Eliot neglects to mention that it features the kind of complicated disguise that he professed to abhor – Freeman was certainly very keen on keeping the genre pure. In a 1924 essay on 'The Art of the Detective Story', Freeman complained that too few readers and critics had learned to recognize clearly the distinction between serious detective fiction and the 'crude and pungent sensationalism' of the vulgar thriller, in which 'the writer's object is to make the reader's flesh creep.' Especially deplorable was the way in which the thriller writer sought to induce a rising curve of stimulation, delivered at a constantly accelerating pace. There was invariably a 'vast amount of rushing to and fro of detectives or unofficial investigators in motor cars, aeroplanes, or motor boats, with a liberal display of revolvers or automatic pistols and a succession of hair-raising adventures.' In sharp contrast, the 'intellectual satisfaction' found in 'good detective fiction' required 'the power of logical analysis and subtle and acute reasoning', a 'quality which is the most difficult to attain, and which costs more than any other in care and labour to the author.' Freeman firmly believed that the writer of the thriller faced no such problems. Here was a profligate mode of writing, untroubled by its own excessive productivity and perfectly in synch with the new and morally dangerous world of the cinema, mindlessly flooding the market. Worse still, the thriller set in train a vicious spiral, for as the reader becomes inured to shock, 'the violence of the means has to be progressively increased in proportion to the insensitiveness of the subject.'

Freeman thought it self-evident that 'a form of literature which arouses the enthusiasm of men of intellect and culture' was immune from being affected by any 'inherently base quality'.[16] The detective story therefore displayed a perfect symmetry of audience and genre. Predictably, this stratification of the intellect was carried over into Freeman's own fiction. In the first of his extremely popular Dr Thorndyke books, *The Red Thumb Mark* (1907), it is plain that the most horrifying experience of the falsely accused (invariably middle class) is to be subjected to the extremes of downward social mobility. 'The law professes to regard the unconvicted man as innocent; but how does it treat him?' Dr Thorndyke's grim answer to his own question is worth quoting:

> He will be ordered about by warders, will have a number label fastened on to his coat, he will be locked in a cell with a spy-hole in the door, through which any passing stranger may watch him; his food will be handed to him in a tin pan with a tin knife and spoon; and he will be periodically called out of his cell and driven round the exercise yard with a mob composed, for the most part, of the sweepings of the London slums.

These observations are given added poignancy by the facts of the narrator's own situation: Jervis, Thorndyke's Watson, is an unemployed doctor whose studies have forced him into debt and, as if in counterpoint to his dismal circumstances, fear of contamination by the feckless and the poor looms large in the book. Even the majesty of the high court seems to have been tarnished by constant contact with the great unwashed: 'plain and mean to the point of sordidness', its woodwork is 'poor, thinly disguised by yellow graining, and slimy with dirt wherever a dirty hand could reach it.'[17]

These anxieties are given a somewhat more sinister twist in another of Freeman's ventures into cultural criticism written three years before his essay on 'The Art of the Detective Story'. Wearing his other professional hat as a qualified physician, Freeman had published a book on eugenics entitled *Social Decay and Regeneration* (1921) in which he attempted to isolate the main causes of what he saw as Britain's precipitous decline. Chief among these were a drift towards collectivism as a result of World War I, a development which was interrupting 'the continuity of social evolution'; and the spread of mechanized production methods.[18] In both cases the effects are fundamentally the same: 'a sheep-like submissiveness' (p. 31) and a general lowering of intelligence due to the 'mental exhaustion produced by the strenuous but intolerably dull and monotonous labour associated with factory work' (p. 192). Throughout the nation one could discern 'a diminished clarity and precision of thought, a tendency to confused thinking and a failure to distinguish between the essential and the subsidiary' (pp. 268–9) – precisely those defects which he was shortly to identify with the thriller.

The terms of Freeman's eugenics exactly mirror those of his appraisal of detective fiction – indeed, I want to argue that they are one and the same. The world is being engulfed by a tidal wave of mediocrity, a slew of inferior commodities and inferior people, and this process of degeneration is all the more catastrophic because of the speed with which it is happening. The category of the socially 'unfit' is expanding at an alarming rate: not only among the 'abnormal' unfit, that 'large class of degenerates' who make up the bulk of the criminal strata (p. 243), but also among those 'comprehensively inferior' human beings that Freeman dubs the 'normal' unfit. This latter group accounts for approximately one-fifth of the population and is 'the most prolific class in the whole community'. Its most worthless specimens are comparable to 'the African negro', though their lack of primitive vigour makes them 'dull', 'helpless and unhandy' (p. 250). Like Chesterton, Freeman was also concerned about the Jews, 'the alien sub-man, diffusing racial as well as personal inferiority' throughout the body politic, though his primary focus was upon the poorer immigrant rather than the 'plutocrat' (p. 267).

The solution to this calamity advanced in *Social Decay and Regeneration* was a stern reassertion of boundaries, a new political geography based upon 'the voluntary segregation of the fit'. It was necessary to create 'visible aggregations' of 'men and women of good racial quality' (p. 314) and these clusters of healthy, intelligent families living side by side and interbreeding would have a powerful 'moral effect' on the rest of the community, renewing civil society by reawakening its 'racial conscience' (p. 310) – interestingly, one of the models for these so-called 'Lodges' was 'the Freemasons' (p. 314). Transposed to the plane of detective fiction this argument follows a parallel logic: in asserting the integrity of the detective story genre, Freeman hoped to reveal the close link between 'rigid demonstration' and 'artistic effect', thus helping to undermine the specious attractions of the thriller. But population imbalance also had implications for the future of the detective story. There were simply not enough of the right kinds of readers, that is 'men of the definitely intellectual class: theologians, scholars, lawyers and to a lesser extent, perhaps, doctors and men of science.'[19]

I am not claiming that eugenics is the logical culmination, nor even an inherent predisposition, among the more high-minded representatives of the Golden Age. Chesterton, notwithstanding his sometimes virulent antisemitism, was an honourable campaigner against eugenics – and it should also be remembered that he spoke out against Jewish persecution towards the end of his life. Dorothy L. Sayers was another writer who sought in her criticism to distinguish firmly between 'Intellectual' and 'Sensational' modes of writing, yet her penultimate novel *Gaudy Night* (1935) contains a mercilessly satirical portrait of a eugenicist, though her barbs are reassuringly directed at the figure of an American, Miss Schuster-Slatt.[20] But while these examples suggest that the anxieties surrounding detective fiction in this period were extra-generic in origin, it is remarkable how easily discussion of fictional matters slips into a generalized language of crisis, as if the state of the detective story could serve as an index of the health of the nation.

Throughout the Golden Age there is a constant sense that, despite all its manifest achievements, detective fiction remains a thoroughly embattled genre, uncertain whether it can continue to reproduce itself successfully or even manage to stay alive. In short, there is always the lurking fear that the detective story will soon become another corpse in the library. This fear takes two basic forms. On the one hand, writers and critics were afraid of a slide into entropy, anticipating that what *John O'London's Weekly* called 'the well-reaped field of detective fiction' had already been over-cultivated and was rapidly reaching a state of utter depletion; or, to shift metaphors, that the bubble was about to burst.[21] At the same time, however, because the detective story was *internally* weak, it was also perceived as being vulnerable *externally*, in danger of being absorbed into its less principled competitors. It was as though, after once having strenuously separated itself from its lowly provenance in melodrama and the Gothic, detective fiction was now on the verge of collapsing back into them, of sinking down into the netherworld of the subliterary. One might say that these fears of depletion and contamination, rebarbatively literalized, are precisely the fears that so disquietingly animate the pages of *The Man Who Knew Too Much*. Except that these fears switch registers so readily that they are hardly distinguishable from each other, moving from the particular to the general in the blink of an eye, like the reversible sides of a single garment.

This is why the Golden Age's own self-validating rhetoric is so apprehensive, so nervous, fretting endlessly about its readership, about its literary constituency. Because crime writing really did dominate popular literature during the interwar years, the ambitions and the anxieties that are cathected on to it were immense. One can see both sides of the coin in Dorothy L. Sayers' work. Writing in *The London Mercury* in November 1930,[22] she observed that 'something odd' had happened to the mystery story: it had moved upmarket and, rather than being read solely 'in back-kitchens', was 'becoming more and more high-brow in its appeal.' In Sayers' cautious appraisal, however, this apparent success seemed to contain the seeds of its own downfall: for while it was appreciably 'more subtle, literary and desiccated in manner', it was also 'in great danger of losing touch with the common man, and becoming a caviare banquet for the cultured.'

Sayers dreamed of ever higher peaks of literary achievement, but it is instructive that, in spite of her enormous *élan*, she too frames these hopes in terms of a vocabulary of crisis. Her major concern was whether the detective story had any real 'chance of surviving… as part of the main stream of a nation's literature', and she was evidently worried that the genre had become too narrow to combine 'fine writing with common feeling'. Behind her valorization of 'fine artistic seriousness' lies a deeply-held and still more serious desire that detective fiction might begin to reunite the entire nation, healing what, in a peculiarly homely (not to say, *gendered*) metaphor, she imagines as the old divisions between 'the kitchen' and 'the study'. It is the cultural insularity of the mystery story that especially troubles Sayers, her 'fear' that, 'in abandoning the crude vulgarities of the past', it 'may have become the literature only of a single highly-sophisticated and over-sensitive class.'

From this standpoint, Sayers' *Gaudy Night* can be read as a cautionary tale of what can happen when the proper relationship between the kitchen and the study breaks down. Sayers' own letters from late 1935 when the book was about to go into production show her fierce commitment to the defence of intellectual integrity, *Gaudy Night*'s central theme.[23] Yet what makes the novel such an extraordinary performance is not only the care with which class and gender are folded into the narrative, but Sayers' skilful (though nonetheless highly didactic) orchestration of the wider political issues with which she was concerned. There is a seamless web connecting the proto-fascism of the college servants in the book's Oxford setting to Lord Peter Wimsey's mysterious off-stage diplomatic negotiations, and from the naive enthusiasm for eugenics to the burning of scholarly books and manuscripts. At the same time, it is immensely important that the violence that is unleashed against the women dons in Harriet Vane's old college is rooted in the failings of the servant class when its members strive to rise above their appropriate station in life. In seeking revenge for her husband's suicide after he has been exposed for falsifying his academic work, Annie Wilson, the guilty college 'scout', shows that she has been acting out of ignorance, the ignorance of her class. Because she had married a man who was a cut above her socially, she had never understood the meaning of the scholar's vocation and its unwavering dedication to the pursuit of truth. She cannot see the college as a repository of timeless values; to her it is merely a place 'where you teach women to take men's jobs and rob them first and kill them afterwards', where 'an old book or bit of writing' matters more than people.[24]

In a novel whose main protagonists are painfully struggling to find a satisfactory balance between reason and emotion, mind and body, this wounding outburst of raw emotion from an uneducated woman serves as a warning of the dangers of blind devotion. But Annie Wilson's accusation is also aligned with Nazi ideology, since she is willing to sacrifice the truth to *ressentiment*, believing that a woman's place is in the home, supporting her man, no matter what. By contrast, the relationship between Lord Peter Wimsey and Harriet Vane offers the possibility of a very different kind of cross-class marriage, companionate *and* egalitarian, domestic *and* intellectual. What this loaded comparison obscures, of course, is the uncomfortable truth that in this period fascist sentiments were far from being the exclusive preserve of working-class men and women or foreigners.

Sayers was not the only Golden Age writer to explore the politics of domesticity. When Margaret and G.D.H. Cole presented their hero Superintendent Wilson to the public as part of a radio series on crime fiction entitled *Meet the Detective* (1935) they depicted him as 'educated, but not highbrow', a man of exemplary ordinariness.[25] Unlike Sherlock Holmes or Dr Thorndyke, Wilson's success as a sleuth was based upon a combination of common sense and sexual equality:

> You consult your wife, like a sensible husband, when you get stuck. You trust to reason, with a bit of luck thrown in, and don't talk

> rubbish about the higher psychology and that sort of thing. In fact, you behave as the ordinary intelligent man does, with the advantage of good training and an organisation behind you and a bit more sense than average. (p. 115)

This unassuming, but highly effective embodiment of rationality is more than just a good detective. He is also the representative new man of the Coles' reformist socialist imaginary, a figure with whom the man and woman in the street can identify, 'the sort of detective who enables the reader to put himself in your place' (p. 117). Revealingly his biography contains an impeccable mixture of talent and disadvantage: born of 'respectable but impecunious, middle class parents', Wilson won 'a scholarship to a secondary school' and chose police work over office work, an alienated lifetime spent 'sitting on a stool and entering things in a ledger' (p. 116).

What one sees again and again in Golden Age detective fiction is an attempt, in some cases discreetly muted, in others glaringly explicit, to articulate a vision of social reconstruction (or what Freeman called 'social regeneration') in which the desire to conjure up a new reading public becomes part of a wider project of imagining a whole new set of relationships between men, women, and social classes. This project – in reality, a series of disparate projects – is best understood as a species of *populism*. In their very various ways each of the writers discussed in this essay sought to define and occupy a strategically central location (not necessarily or not always a 'middle ground') from which a range of social identities could be fused into an ideological alliance whose adherents might be said to represent an authentic core of public opinion. Put more simply, populism involves imagining a people in order to claim to speak on their behalf. Much of the seriousness of Golden Age fiction and criticism ultimately derives from this concern with social leadership and opinion formation.

Chesterton's populism is widely acknowledged.[26] But what of a writer like Sayers, so often scorned today for her snobbery and her fascination with privilege? Is not her sturdy defence of the study the very antithesis of populism? Not quite. Looking back over her career as a detective novelist, in 1941 at the close of the Golden Age, Sayers conceded that 'ideas can't be violently imposed on people', but then confessed that she had tried as hard as she could 'to "infiltrate" a few general ideas' into books like *Gaudy Night*, as 'naturally' and as 'carefully' as possible. But now she wanted to reach out to 'the mass of the nation' using pamphlets and speeches, beginning by 'tackling the people nearest to one's self.'[27] It is sometimes misleadingly suggested that the reason Sayers abandoned the detective story was that she fell in love with her own hero and that the marriage of mystery and romance led her into a creative impasse. But a far more likely explanation is that Sayers had become thoroughly enamoured of her rôle as cultural critic and that the power of its attraction was too strong to resist, a sure sign that the Golden Age really was coming to an end. After 1938 she started to channel her energies into other media like the theatre, radio drama, and the paperback classic, exchanging Wimsey for Dante, and becoming a much more public intellectual in the process. One of Sayers' least remembered books was prompted by the onset of World War II, a short guide to social reconstruction of her own, urgently entitled *Begin Here* (1940). Though they have long since vanished from popular memory, Sayers' final remarks might stand as a credo for all the writers I have been discussing in this essay. 'Words ... are real and vital', she solemnly assured her readers. 'They are a form of action' and 'they can change the face of the world.'[28]

Notes

1 Dorothy L. Sayers to E.C. Bentley, 17 April 1936, in *The Letters of Dorothy L. Sayers*, Vol. I, *1899–1936: The Making of a Detective Novelist*, ed. Barbara Reynolds (London: Hodder & Stoughton, 1995), pp. 387–8.
2 Howard Haycraft, *Murder for Pleasure: The Life and Times of the Detective Story* (New York and London: D. Appleton-Century, 1941), p. 158.
3 On the 'laws' governing detective fiction, see S.S. Van Dine, 'Twenty Rules for Writing Detective Stories' (September 1928) repr. in Howard Haycraft (ed.), *The Art of the Mystery Story* (1946; repr. New York: Grosset & Dunlap, 1961), pp. 189–93.
4 Anthony Berkeley's phrase is taken from the preface to his novel *The Second Shot* (1930), quoted by Julian Symons, *Bloody Murder: From the Detective Story to the Crime Novel* (Harmondsworth: Penguin, 1985), p. 107; Dorothy L. Sayers' words are from her letter to E.C. Bentley, *Letters*, p. 387. Both writers were founder members of the Detection Club who became critical of what they saw as the limitations of the genre.
5 T.S. Eliot, 'Books of the Quarter', *The New Criterion,* 5.1 (January 1927), 139–43.
6 Ronald A. Knox, 'A Detective Story Decalogue', (1928) repr. in Haycraft (ed.), *The Art of the Mystery Story*, pp. 194–6.
7 Eliot, 'Books of the Quarter', pp. 141–2.
8 Ibid., pp. 140–1.
9 Uri Eisenzweig, *Le Récit impossible: forme et sens du roman policier* (Paris: Christian Bourgeois Editeur, 1986), p. 10, my translation.
10 Eliot, 'Books of the Quarter', p. 141.
11 G.K. Chesterton, *The Man Who Knew Too Much* (1922; repr. New York: Carroll & Graf, 1989), p. 9. Subsequent page references are given in parentheses in the text.

12 John Buchan, *The Thirty-Nine Steps* (1915; repr. London: Pan, 1947), p. 12.
13 'Why I like Detective Stories', in Gertrude Stein, *How Writing is Written*, ed. Robert Bartlett Haas (Los Angeles: Black Sparrow Press, 1974), p. 150.
14 See Anthony Julius, *T.S. Eliot, Anti-Semitism and Literary Form* (Cambridge: Cambridge University Press, 1995).
15 Eliot, 'Books of the Quarter', p. 143.
16 R. Austin Freeman, 'The Art of the Detective Story' (1924) repr. in Haycraft (ed.), *The Art of the Mystery Story*, pp. 7–17.
17 R. Austin Freeman, *The Red Thumb Mark* (1907; repr. New York: Carroll & Graf, 1986), pp. 92, 212.
18 R. Austin Freeman, *Social Decay and Regeneration* (London: Constable, 1921), p. xi. Subsequent references are given in parentheses in the text.
19 R. Austin Freeman, 'The Art of the Detective Story', pp. 11 and 16.
20 On Sayers' distinction between the 'Intellectual' and the 'Sensational', see her Introduction to her collection of *Great Short Stories of Detection, Mystery, and Horror* (London: Gollancz, 1928).
21 The phrase occurs in a review of Dennis Wheatley and J.G. Links' 'murder dossier' *Murder Off Miami*, in *John O'London's Weekly*, 1 August 1936.
22 Dorothy L. Sayers, 'The Present Status of the Mystery Story', *The London Mercury*, 23.133 (November 1930), 47–52.
23 See, for example, Sayers' letters to Muriel St Clare Byrne, 8 September 1935, and to Victor Gollancz, 26 September 1935, in *Letters*, pp. 352–5 and 357–8 respectively.
24 Dorothy L. Sayers, *Gaudy Night* (1935; repr. New York: Harper & Row, 1986), p. 443.
25 G.D.H. and Margaret Cole, 'Meet Superintendent Wilson', in H.C. Bailey *et al.*, *Meet The Detective* (London: George Allen & Unwin, 1935), p. 116. Subsequent references are given in parentheses in the text.
26 See Margaret Canovan, *G.K. Chesterton: Radical Populist* (New York and London: Harcourt Brace Jovanovich, 1977).
27 Dorothy L. Sayers to Mr J. Wilshin, 21 August 1941, in *The Letters of Dorothy L. Sayers*, Vol. II, *1937–1943: From Novelist to Playwright*, ed. Barbara Reynolds (Cambridge: The Dorothy L. Sayers Society, 1997), pp. 285–7.
28 Dorothy L. Sayers, *Begin Here: A War-Time Essay* (London: Gollancz, 1940), p. 156.

5
Sherlock's Children: the Birth of the Series

Martin Priestman

Taking television into account, it is arguable that the two most prevalent fictional forms of our time are the series and the serial. But while the serial or soap opera can be discussed as a variant of the single extended storyline, reflecting its ancestry in Dickens and other Victorian writers of novels that were episodically presented in periodicals, the series demands to be discussed in its own unique terms. Such discussion is still somewhat fitful and quite often the two forms are blurred together, as in the noun 'seriality', which seems to spring from the word 'serial's' double life as a noun and as the adjectival form of 'series'. The series as such is the form which repeats, theoretically *ad infinitum*, the same kind of action in roughly the same narrative space or time-slot, featuring at least one character continuously throughout. Its particular qualities deserve a more accurate name than 'seriality': 'seriesicity', perhaps.

Of course the two forms are often in tension with each other: in particular, the series keeps threatening to develop elements of narrative thrust which if allowed their head would turn it into a serial. In this essay I shall be considering some of the ways in which this perpetual tendency is 'managed' in the work of Arthur Conan Doyle and some contemporaries, and examining some of the implications of this. First, it is perhaps useful to consider the ambiguous term 'serial killer'.[1] It is not at all original to point out that the period of ascendancy of the great series hero Sherlock Holmes coincided with that of Britain's first great sensation over a serial killer, Jack the Ripper.[2] But what differentiated the 'story' of Jack the Ripper from those of the villains featured in the serialised novels of Dickens and others, as well as those of the real-life murderers who had provided staple reading through the pages of the eighteenth-century *Newgate Calendar*, was the lack of narrative closure made inevitable by the failure to identify the Ripper successfully.[3] Almost all the writing generated around Jack the Ripper, then and since, has been partly an attempt to convert the story's maddening 'series' quality into the rounded, reassuring contours of the serial, in which every action is also part of a development leading finally to conclusion. If Jack the Ripper is then the best-known early example of what should be called a series killer, it is not absurd to suggest that one of the great attractions of Sherlock Holmes was his ability to counter this nightmarish 'seriesicity' with his own benign version of the same quality. Where the Ripper evades identification, Holmes establishes it at every turn, but with the same refusal to enter any full-blown narrative that could ever bring his series to an end.

The horror of the series killer is in part that of the anonymity offered by the city: as has often been pointed out, the fictional detective from Poe's Dupin to Holmes and onwards seems tailor-made to calm anxieties about just such urban anonymity. In Poe's proto-detective story 'The Man of the Crowd', it is the utter impossibility of identifying the old man except in terms of his need to stay in the crowd which makes the *flâneur*-narrator pronounce him 'the type and genius of deep crime'.[4] If Poe invented the detective story soon afterwards to suggest that such 'unreadability' could always be surmounted, he also at the same stroke invented the series. Though C. Auguste Dupin appears in only three stories – 'The Murders in the Rue Morgue', 'The Mystery of Marie Roget' and 'The Purloined Letter' – this was enough to establish the basic, highly stable, structure linking the series to fictional detection in particular.

Before continuing to explore this link directly, it will be worth looking briefly at other possible variants of the series. It is arguable that the foundations of Western literature in Homer's *Iliad* and *Odyssey*, and perhaps even the internal structure of the latter, already demonstrate an evolving fascination with the exploits of a single hero. Other legendary heroes such as Heracles and the knights of medieval romance are also celebrated in modes comparable to the series, though as Umberto Eco has argued, each such story 'followed a line of development already established, and it filled in the character's features in a gradual, but definitive, manner.... The story has taken place and can no longer be denied.'[5] It is fundamentally the presumed non-fictionality, the 'pastness', of such heroes that differentiates them from the fictional series hero, forever bound to an endlessly repeatable present.

In his 1962 essay 'The Myth of Superman', Eco explores what he calls the 'iterative scheme' – in fact the series – chiefly in relation to the

Superman comics, but with much broader implications.[6] Contrasting the 'iterative scheme' to the 'self-consuming' nature of classic nineteenth-century narratives and characters, Eco points out first of all the distortion of time the series necessarily involves, characterized by the narrator appearing to pick up 'the strand of the event again and again, as if he had forgotten to say something and wanted to add details to what had already been said'.[7] With Superman, a constant reminder that no 'self-consuming' development must ever occur is provided by the presence of Lois Lane, forever on the brink of discovering Clark Kent's identity with Superman, and hence of destroying the eternal triangle of mutual frustration on which the whole structure depends.

For Eco, this denial of self-consumption is essential to the chief pleasure of the series form: its redundancy of message, its absolute refusal to let any new elements disrupt the repetition of the meaning we could have gleaned from the first episode alone. This is particularly true of the detective story which, despite its apparent promise to 'satisfy the taste for the unforeseen or the sensational is, in fact, read for exactly the opposite reason, as an invitation to that which is taken for granted, familiar, expected'.[8] In socio-political terms, Eco ascribes this taste for redundancy to the prevalence of the opposite in the real life of a contemporary world where economic and industrial change is seismic, cataclysmic and pre-eminently 'self-consuming'. In this it contrasts to the still very stable, settled life against which nineteenth-century bourgeois narrative opposed the aesthetic of developmental, self-consuming change. Since such change is now a reality for all, 'serious' literature echoes this in an aesthetic which only brings out more harshly and self-reflexively its seismic, disruptive quality, while it is left to 'escapist' forms such as the comic and detective story to provide the consolations of eternal repetition and sameness.

As a concluding point, Eco comments on the series form's wilful stifling of the sheer *power* for change its hero inevitably accumulates, or would accumulate if all his achievements were ever to be totted up in sequential 'real time'. We might add that while such power is given for Superman himself, it is a striking and much-noted feature of the detective series that the track record of a Father Brown or a Miss Marple, let alone a Sherlock Holmes, ought progressively to preclude the anonymity on which their success partly depends. This is even more of a problem for police heroes within a supposedly realistic promotion structure: from Michael Innes's John Appleby to P.D. James's Adam Dalgliesh, their inevitable elevation to the top of the force compels a kind of eternal semi-retirement in which new cases as often as not have to be stumbled on in a semi-amateur capacity. For Eco, the denial of such accumulating power for change, and its constant rechannelling back into a series of small, discrete cases revolving round the private restitution of private property, have a precise political significance. The 'need to forbid the release of excessive and irretrievable developments' demands that 'Superman *must* make virtue consist of many little activities on a small scale, never achieving total awareness'.[9]

If the use Superman *could* make of his power would still be only fictional, Eco suggests that it could 'furnish us in the meantime with a definition that through fantastic amplification could clarify precise ethical lines everywhere' and perhaps, Eco implies, bring forward a social revolution in the process.[10] One fictional figure who actually did so, Eco argues in a later essay, is the Prince Rodolphe of Eugène Sue's *Les Mystères de Paris* (1844) whose exploits among the poor of Paris helped to fuel the Revolution of 1848, for all Marx's and Engel's denunciations of him in *Die Heilige Familie* as an essentially paternalistic figure.[11] Significantly for Eco's argument, Sue's sprawling episodic narrative hovers somewhere between series and serial: published in instalments in the *Journal des Débats*, it began as a series featuring Rodolphe as the disguised punisher of gangsters and avenger of isolated wrongs but builds via a kind of dialogue with the public enthusiasm provoked by such episodes into a serial narrative of programmatically interlinked stories revealing particular class-figures as the oppressors of the proletariat. What might seem the least socially 'progressive' element of the series/serial, the hero's royal status, in fact functions simply to provide the superman-like power of 'fantastic amplification' already mentioned.

Halfway between Sue's 1840s and Superman's heyday in the 1940s comes the Sherlock Holmes of the 1880s and 1890s. Recent criticism, such as Stephen Knight's and Dennis Porter's, has often dealt superbly with the effectiveness of this figure in reinforcing conventional bourgeois stereotypes.[12] Franco Moretti's 'Clues', from his *Signs Taken for Wonders* (1983), encapsulates many recurrent motifs of such criticism: the 'singular' clue as a sign of suspicious individuality in an age when entrepreneurship has given way to corporate capitalism; the complicity of criminal and victim in anti-bourgeois forms of 'upstart' or 'aristocratic' greed; the role of history as an irruptive 'violation' of an ideally eventless present; the famous image of Holmes flying over London and 'gently removing the roofs' to 'peep in' like the unseen seer of Foucault's image of the Benthamite panopticon prison; the more general resemblance of Holmes's practice to those of the two medical doctors, Doyle and Watson, who invent and narrate it – again seen in Foucauldian terms of totalizing social control.[13]

Many of these points can be related closely to the series mode. The 'singular' clue or crime is ironically both the *sine qua non* required to distinguish every episode from every other, and the phenomenon whose larger claim to independent meaning the series as a whole seems precisely designed to crush. The dangerous impetus of history from past to future is precisely dissipated in a mode dedicated to the faith in an endlessly repeating present. The Foucauldian images of capillary panopticon-like or quasi-medical surveillance merge perfectly with the mode's endless enactment of a curiosity over human life precisely defined by its divisibility into a listable, tabulatable string of 'cases'.

What I would like to follow up now relates to the series form's particular pleasures, and especially that of its 'management' of its own seriesicity. Most of the rules here were established by Doyle, who also definitively described the form's economic basis in his autobiography *Memories and Adventures*: 'It had struck me that a single character running through a series, if it only engaged the attention of the reader, would bind that reader to that particular magazine.'[14] More than Poe with his three irregularly-produced Dupin stories, or the unfairly neglected Emile Gaboriau, whose novel-series featuring the detectives Tabaret and Lecoq Doyle imitated in *A Study in Scarlet* and *The Sign of Four*, the Doyle who hitched his occasional and not very successful hero Sherlock Holmes to the fortunes of the *Strand* magazine in 1891 was the real inventor of the series as we now know it.[15]

Thereafter, famously, comes the growing struggle between the would-be serious historical novelist and the monster he has spawned, culminating in that other struggle above the Reichenbach Falls between Moriarty, the brilliant intellectual who can create his own plots, and Holmes, the maddeningly repetitious creature who can only frustrate those of others.[16] If the only way of killing Holmes involves the artist's demise too, or at least that of a large part of his income, so be it – except of course that after much public mourning in the streets and a hefty raise from the *Strand*, so it wasn't. The public outcry over Holmes's death was the first great – and somewhat frightening – expression of the emotional dependence fostered by the series mode. For all that might be said about the uniqueness of Holmes's personality or Doyle's brilliance as a writer, it was surely the raw power of seriesicity itself, flexing its emotional and financial muscles for the first time, that produced the outcry.

In fact, the notion of Holmes as a unique or superbly characterised personality is arguably a myth: what is far more interesting about him is the way in which he encapsulates seriesicity itself within a fairly loose envelope of potentially contradictory traits. These traits are initially scraped together from other sources: from Poe's Dupin, as is generally recognized, but also from both Gaboriau's series heroes Tabaret and Lecoq – one a forlorn and frustrated bachelor and the other a ferret-eyed policeman on the make. From the more energetic, animal-like elements of these characters emerge the brash, anti-intellectual Holmes of *A Study in Scarlet*, whose sneers at Dupin and Lecoq as detectives hold an honourable place in the oedipal predecessor-bashing which is one of the ritual pleasures of series detection.[17] From the more reflective, melancholy sides of the same characters emerges the completely different Holmes of *The Sign of Four*, an intellectual aesthete to his fingertips whose drug-induced fondness for literary quotation and bored distantiation from his friend Watson's gruff heterosexuality may have been influenced by a meeting with Oscar Wilde, whose *Dorian Gray* appeared in the same magazine (*Lippincott's*, not yet *The Strand*) in 1890.[18]

By various devices, including a repeated leaping back and forth in time to periods before and after the series's one irreversible event – Watson's marriage in *The Sign of Four* – the series proper manages to blend, or blur, these two completely different Holmeses into an apparent unity. As I have argued elsewhere, this 'dual nature' – born originally from accident, or the carelessness of writing without continuation as a series in mind – is increasingly interpretable as an expression of the tensions within the series mode itself.[19] The decisiveness, the refusal of large ideas, the need to get to the point, express the brevity of the short story and the reader's need to emerge with a single apex-like solution to all that has been said and done. On the other hand the dreaminess, the quest for something ever-more 'singular' and 'outré', the growing 'ennui' and dread of routine, express the series form's growing ability and urge to reflect on its own repetitiousness. The very early story 'The Red-headed League', in which Holmes's 'dual nature' is most explicitly discussed, is also the one where many images of repetition and urban reproducibility are offset with the ability of two individuals – Holmes and the criminal genius John Clay – to think beyond and through these, to great comic effect. The fantastic newspaper ad attracting red-headed men from all quarters of London, the pointless labour of writing out the *Encyclopaedia Britannica* from A to Z, the 'manufactory of artificial kneecaps' where the crooks' trail runs cold, are offset with the knowledge of the real underlying connections which make all these absurd phenomena useful and interpretable to Clay and Holmes alone.[20] This awareness of the dance between reproducibility and uniqueness is, arguably, what makes 'Sherlock Holmes' tick, both as a series and as a man.

I would like now very briefly to turn to some of Doyle's contemporaries. Though G.K. Chesterton's Father Brown series began 20 years later in 1911, Holmes was still going strong at this time and had set up the genre very clearly by then as something available to be commented on. Chesterton's commentary begins in the first Father Brown story 'The Blue Cross' with a thought by the apparent detective-hero Valentin that 'the criminal is the creative artist; the detective only the critic'.[21] The two apparent criminals he is trailing at this time are the real super-thief Flambeau and the 'little' priest Brown, who relieves Flambeau of his stolen treasure after successfully 'criticising' his flawed theology, in time for Valentin to make the final dramatic pounce. In the next story, 'The Secret Garden', Valentin apparently 'creates' a superb scenario with potential victims and suspects all gathered in a classic 'locked space', in which the unassuming Brown again unveils the criminal, but this time the crime is a murder committed by Valentin himself. In the next two stories Brown again crosses swords with the flamboyant Flambeau, so successfully that thereafter Flambeau repents and becomes in effect Brown's Watson for the rest of the long, long series.

Elsewhere, somewhat earlier, Arthur Morrison's series detective Dorrington (1897) performs similarly impressive feats of solo deduction, but only in order to rob and even murder those he investigates, purely for financial gain. In creating such a figure, the socialist Morrison goes far beyond comparable series rogues Arsène Lupin and Raffles, all perhaps distant relatives of Sue's masked avenger Prince Rodolphe, in what amounts to a breathtaking challenge to the loyalty evoked by seriesicity itself. A thin but subversive line of such series villains can be traced from Dorrington through Marcel Allain and Pierre Souvestre's *Fantômas* series (beginning 1911) and Patricia Highsmith's audacious run of novels featuring the murderous Ripley (beginning 1957).[22]

Another *fin-de-siècle* socialist writer, Israel Zangwill, plays with the idea of the extended detective series in *The Big Bow Mystery*, published serially in the *Star* newspaper in 1891, whose celebrated detective Grodman turns out to have committed the murder purely to put the finishing touch to the long series of 'true-crime' narrations of his triumphs penned by a pathetic hack. It is made clear that the detective, the hack and the public who lap up his adventures are all egging each other on in an unholy alliance to produce the 'perfect murder' which eventually transpires.[23] Without space to draw the ironies of all these texts together, I would simply argue that – as with *Tristram Shandy* and the bourgeois novel – for the most penetrating critique of what has been enshrined as a 'natural' and inevitable narrative form, one needs to look closely at the period of its origins. By examining the detective writing of the 1890s, 1900s and 1910s, we can begin to unpick the 'seriesicity' which is so much an assumed fact of our contemporary life that we have never quite found a word for it.

To conclude in the present: my references to Eco, Foucault and Moretti have, if anything, presented the case against the series, in language borrowed from the 1960s and 1970s – more earnestly critical decades than this one, in which it was believed we had more choice about standing outside such compelling cultural structures than perhaps now seems possible. To return to the genre's division of experience into a string of 'cases' as embodying the kind of panopticon-like or quasi-medical surveillance explored by Foucault: television drama, in Britain at any rate, is now a simple choice between various forms of serial and the series format which is itself almost solely a choice between police detection and medicine. What the series form requires is, as we have seen, a combination of two axes: the unique, personal-interest drama of the individual episode and the reassuring repetition of the continuum. Hence the ideal, almost the required, heroes of the continuum are professionals whose job consists entirely of processing people who are themselves at a point of crisis in their lives. Despite sporadic attempts to identify other such figures, police detectives, doctors and nurses clearly come top of the list here, with only lawyers, forensic scientists and criminal psychologists (as in *Cracker*) running them at all close. And we may notice that these are also all the professions Foucault links most closely to the capillary mechanisms of social control, exerting power through an appallingly intimate intensity of knowledge.[24] Through the series mode, we the public come to understand and 'own' this practice of total surveillance, apparently so clearly carried out on our behalf.

Notes

1 As an illustration of the kind of ambiguity I mean: Patricia Cornwell has recently exploited a current obsession with serial killers, not only by tracking

individual ones down in successive books in her Kay Scarpetta series, but also by allowing a single such killer, Temple Gault, to escape serially, as it were, through a sequence of five until finally tracked down in *From Potter's Field* (New York: Scribner, 1995).

2 See Derek Longhurst, 'Sherlock Holmes: Adventures of an English Gentleman 1887–1894', in Derek Longhurst (ed.), *Gender, Genre and Narrative Pleasure* (London: Unwin Hyman, 1989), p. 65.

3 For one of many versions of the *Newgate Calendar* narratives, see George Theodore Wilkinson, *The Newgate Calendar*, ed. Christopher Hibbert (London: Cardinal, 1991). For an excellent account of the dynamics of such narratives, see Stephen Knight, *Form and Ideology in Crime Fiction* (London: Macmillan, 1980), Chapter 1.

4 'The Man of the Crowd', *The Complete Tales and Poems of Edgar Allan Poe* (New York: Random House, 1938), p. 481. See Dana Brand, 'From the *Flâneur* to the Detective: Interpreting the City of Poe', in Tony Bennett (ed.), *Popular Fiction: Technology, Ideology, Production, Reading* (London and New York: Routledge, 1990), pp. 220–37.

5 Umberto Eco, 'The Myth of Superman', in *The Role of the Reader: Explorations in the Semiotics of Texts* (London: Hutchinson, 1981), p. 108.

6 Ibid., p. 117.

7 Ibid., p. 114.

8 Ibid., p. 120.

9 Ibid., p. 124.

10 Ibid., p. 123.

11 Umberto Eco, 'Rhetoric and Ideology in Sue's *Les Mystères de Paris*', in *The Role of the Reader*, pp. 129–30. See also Karl Marx and Friedrich Engels, *The Holy Family, or Critique of Critical Critique* (London: Lawrence & Wishart, 1956), esp. pp. 217–75.

12 Stephen Knight, *Form and Ideology in Crime Fiction*, and Dennis Porter, *The Pursuit of Crime: Art and Ideology in Crime Fiction* (New Haven: Yale University Press, 1981).

13 Franco Moretti, 'Clues', in *Signs Taken for Wonders: Essays in the Sociology of Literary Forms*, revised edn (London and New York: Verso, 1983). See also Arthur Conan Doyle, 'A Case of Identity', in *The Penguin Complete Adventures of Sherlock Holmes* (London: Penguin, 1981), p. 191, and Michel Foucault, *Discipline and Punish: The Birth of the Prison* (London: Allen Lane, 1977).

14 Arthur Conan Doyle, *Memories and Adventures* (London: Hodder & Stoughton, 1924), p. 95.

15 See Emile Gaboriau, *L'Affaire Lerouge* (1866) and *Monsieur Lecoq* (1869) for the fullest portraits of Tabaret and Lecoq respectively.

16 See 'The Final Problem' in *The Penguin Complete Adventures of Sherlock Holmes*, pp. 479–80.

17 See *A Study in Scarlet*, in ibid., pp. 24–5.

18 See Owen Dudley Edwards, *The Quest for Sherlock Holmes* (Harmondsworth: Penguin, 1984), p. 15.

19 Martin Priestman, *Detective Fiction and Literature: The Figure on the Carpet* (Basingstoke and London: Macmillan, 1990), p. 86. Since this paper was delivered, the current paragraph and the preceding one have appeared in

my *Crime Fiction from Poe to the Present* (Plymouth: Northcote House, 1998), pp. 14–15.

20 See Priestman, *Detective Fiction and Literature*, pp. 86–90; and 'The Red-headed League', *The Penguin Complete Adventures of Sherlock Holmes*, p. 182.

21 G.K. Chesterton, *The Innocence of Father Brown* (Harmondsworth: Penguin, 1950), p. 12.

22 See Arthur Morrison, *The Dorrington Deed-Box* (1897); Maurice Leblanc, *Arsène Lupin, Gentleman Cambrioleur* (1907); E.W. Hornung, *Raffles the Amateur Cracksman* (1899); Marcel Allain and Pierre Souvestre, *Fantômas* (1911, English translation London: Pan, 1987); Patricia Highsmith, *The Talented Mr Ripley* (1957), *Ripley Under Ground* (1970), *Ripley's Game* (1974), *The Boy Who Followed Ripley* (1980). For the *Fantômas* series, see chapter 13 in the present volume by Robert Vilain.

23 To be found in *Three Victorian Detective Novels*, ed. E.F. Bleiler (New York: Dover, 1978), pp. 199–302.

24 See Michel Foucault, *Madness and Civilization: A History of Insanity in the Age of Reason* (London: Tavistock Publications, 1967); *The Birth of the Clinic* (London and New York: Pantheon, 1973); *The History of Sexuality* (London: Allen Lane, 1979).

6

Making the Dead Speak: Spiritualism and Detective Fiction

Chris Willis

The idea of any link between spiritualism and detective fiction seems totally contradictory. Classic detective fiction is a literature of logic in which everything has a scientific explanation. It is concerned with hard facts and encourages scepticism. The reader must learn to doubt everything he or she is told about events and characters and must automatically disbelieve such things as alibis. Spiritualism, on the other hand, involves suspension of logical faculties to believe in events and phenomena which cannot be explained in scientific or logical terms. However, it is interesting to note that the rise of the fictional detective coincided with the rise of spiritualism. Both began in the mid-nineteenth century and were widely popular in Britain from the turn of the century until the 1930s. Both attempt to explain mysteries. The medium's rôle can be seen as being similar to that of a detective in a murder case. Both are trying to make the dead speak in order to reveal a truth.

A murder mystery could be solved beyond doubt if the victim could return from the grave to name the murderer. There was a medieval belief that a corpse's blood would flow if the murderer touched it, a superstition which is used to good effect to frighten a suspect into confessing in Ellis Peters' first Brother Cadfael story, *A Morbid Taste for Bones* (1977). It is widely believed that detectives investigating the Jack the Ripper murders opened the eyelids of at least one victim in the belief that the victim's eyes might somehow have retained the image of the last thing she saw. The ability to make the dead communicate in a more reliable way would no doubt be a great advantage for a detective. In Peter Lovesey's historical detective novel *A Case of Spirits* (1975), an ardent female Spiritualist tries to convince the detective that he has mediumistic qualities. She tells him, 'It doesn't prevent you from being a detective as well, you know. I should think it would be a positive advantage.'[1] At the end of the book, the detective needs no supernatural powers whatsoever to reconstruct the seance at which the murder took place and to reproduce and explain the tricks used by the fraudulent medium.

The best-known link between spiritualism and detective fiction is of course Sir Arthur Conan Doyle. As a young man, Conan Doyle had described spiritualism as 'the greatest nonsense upon earth',[2] but in later years he became one of its staunchest advocates. His conversion began in the mid-1880s, when he began to attend seances with friends. Some sessions produced bizarre results: in one a spirit named Dorothy Postlethwaite told him that there was life on Mars.[3] Conan Doyle was initially 'very critical as to the whole proceedings' but, feeling that spiritualism presented 'a problem to be solved',[4] he began to investigate further.

Tragedy struck Conan Doyle's family in 1893 when his wife Louise developed incurable tuberculosis (she died in 1906). In October 1893, Conan Doyle's father died. Biographers point out that this personal loss was echoed in Conan Doyle's work: Sherlock Holmes 'died' at the Reichenbach Falls two months after the death of Conan Doyle's father.[5] Three weeks after his father's death, Conan Doyle joined the Society for Psychical Research.[6] Throughout the 1890s and 1900s, he continued to study what he later described as 'the wonderful literature of psychic science and experience'.[7] His interest in the supernatural found expression in his fiction of the 1890s and 1900s. Perhaps the best-known example is his 1899 ghost story 'The Brown Hand', narrated by a doctor who is also a member of the Society for Psychical Research. A surgeon is haunted by the ghost of a man whose hand he amputated many years earlier: the patient believed in bodily resurrection, and his ghost keeps returning to look for the missing part of his body. The psychic investigator gets rid of the ghost by providing him with a 'new' hand amputated from another patient.[8] In another short story, 'Playing with Fire', Conan Doyle gives a humorous account of a fictional seance at which a unicorn materialises and rampages through the building, much to the consternation of everyone present.[9]

Conan Doyle's full conversion to spiritualism came during World War I. In 1907 he had married Jean Leckie and during the war the couple shared their house with Jean's friend Lily Loder-Symonds, a keen spiritualist. In 1916 Loder-Symonds claimed to have received spirit messages from her brothers who had died early in the war, and from Jean's brother Malcolm, who had been killed at the battle of Mons.

After this, Conan Doyle gradually became convinced of the truth of spiritualism. His conversion was closely linked with personal tragedy. His son, brother and brother-in-law all had been killed in the war. Like many others bereaved in World War I, he found that the spiritualist revival of the 1920s seemed to offer a means of maintaining contact with loved ones after their death.

Conan Doyle lectured widely on spiritualism and in 1922 undertook a tour of America. Despite his unwavering belief in mediumship, he exposed two fraudulent mediums, twins Eva and William Thompson, at a seance in New York. Conan Doyle may have had his suspicions aroused by a strange coincidence: one of the participants at the seance had the surname Moriarty.[10] During this tour, Conan Doyle arranged a private seance for his friend Houdini, at which Jean, acting as medium, produced a 'fifteen page letter from [Houdini's dead] mother.'[11] Houdini was sceptical about the letter, as it was in English and, as he put it, 'although my sainted mother had been in America for almost fifty years, she could not speak, read, nor write English.' Conan Doyle explained this by telling him that she had learnt English in Heaven.[12] Houdini was not convinced.

For the rest of his life, Conan Doyle was a passionate advocate of spiritualism. He wrote several books on the subject, lectured worldwide, and set up a psychic bookshop and library, both of which ran at a loss. In 1901 he had 'resurrected' Holmes; in 1917, he had tried to rid himself of Holmes again – the short story 'His Last Bow' was intended to be the detective's final appearance – but in 1921 Holmes re-appeared in the first of a series of short stories that continued until 1927. There is no doubt that the super-logical, unsuperstitious Holmes brought in the income to subsidise Conan Doyle's spiritualist activities. Summing up his writing career in his autobiography, Conan Doyle hoped that his 'psychic work' would 'remain when all the rest has been forgotten',[13] but the public preferred Holmes. In the mid-1920s, at the height of his spiritualist activities, Conan Doyle had written a short story called 'The Sussex Vampire' in which Holmes provides a natural explanation for an apparently supernatural mystery. Unlike his creator, Holmes disowns any belief in the supernatural, boasting that: 'this Agency stands flat-footed upon the ground, and there it must remain. No ghosts need apply.'[14]

Conan Doyle kept spiritualism out of the Holmes stories, but in several twentieth-century detective stories spiritualist activities, and particularly seances, are used to further the plot. There is an uneasy relationship between detective fiction and the figure of the medium.

After all, the detective would be virtually redundant if the medium could summon a murder victim back from the dead to name the murderer. In traditional fiction, the detective is usually portrayed as a knowledgeable, respected figure, often with a middle-class or upper-class background and a good education. Mediums, on the other hand, are usually portrayed as rather ridiculous figures, almost invariably being ill-educated, badly-dressed, middle-aged and vulgar. In 'Golden Age' fiction, the medium is usually a working-class woman who provides a deliberately incongruous note in stories set among the middle and upper classes. The other characters look down on the medium, but it is not unusual for her to be proved right where they are wrong.

Fraudulent mediums, however, provide easy targets for the detective's abilities. The exposure of a fake medium provides the basic plot for several detective stories, one of the earliest examples probably being Tom Gallon's 1903 short story 'The Spirit of Sarah Keech'[15] in which spirit messages from a dead woman 'miraculously' appear on a typewriter whose keys move even though no one is touching them. In fact the typewriter is connected to another machine in the next room, so that messages typed on one machine appear on the other. Unfortunately for the medium, the typist employed to do this is an undercover detective.

The boom in spiritualism after World War I made the medium a well-known figure in fact and fiction. In 1928 Lilian Wyles, the CID's first woman officer, became involved in a *cause célèbre* when she investigated complaints against a well-known medium. Wyles arranged a private seance, where she was somewhat surprised to hear predictions which referred to her husband and her sister. In fact, Wyles was unmarried and had no sisters. The medium also predicted the imminent death of Wyles' mother – who lived for another 12 years. Not surprisingly, Wyles said that this 'did nothing to impress me as to her powers of clairvoyance'[16] and a summons was issued against the medium, who was fined after a much publicised trial. At that time it was illegal to predict the future and the medium's authenticity was not an issue in the trial. However, Wyles' evidence established that the medium was a fraud. The case caught the public imagination, and the fraudulent or misguided medium was to be a recurring figure in fiction during the years that followed.

Generally speaking, the medium is portrayed as either a person with supernatural powers which no-one quite understands or a fraud who is exposed by the detective. In 'Golden Age' fiction a third situation sometimes arises, when the detective or a confederate pretends to have mediumistic powers to trick someone into a confession or revelation.

Dorothy L. Sayers' *Strong Poison* was published in 1930 at the height of the post-World War I boom in spiritualism. In this novel Lord Peter Wimsey's sidekick, the redoubtable Miss Climpson, masquerades as a medium in order to search a house for a missing will.[17] She holds a series of private seances, during which she produces an impressive array of 'spirit rappings' by means of a small metal soap box attached rather painfully to her leg with a strip of elastic. She also manages an impressive performance of table-turning, making a small bamboo table levitate by supposedly supernatural means. This is done by means of wires attached to her wrists while her hands remain firmly in view on top of the table. Having found out a little about her client's background, she is able to produce convincing 'spirit messages' from a variety of people including her client's dead fiancé. The reader is told that Miss Climpson learned these tricks from:

> a quaint little man from the Psychical Research Society [who] ... was skilled in the investigating of haunted houses and the detection of poltergeists ... she had passed several interesting evenings hearing about the tricks of mediums ... she had learned to turn tables and produce explosive cracking noises; she knew how to examine a pair of sealed slates for the marks of the wedges which let the chalk go in on a long black wire to write spirit messages.
>
> She had seen the ingenious rubber gloves which leave the impression of spirit hands in a bucket of paraffin-wax, and which, when deflated, can be drawn delicately from the hardened wax through a hole narrower than a child's wrist. She even knew theoretically, though she had never tried it, how to hold her hands to be tied behind her back so as to force that first deceptive knot which makes all subsequent knots useless, and how to flit about the room banging tambourines in the twilight in spite of having been tied up in a black cabinet with both fists filled with flour.
>
> Miss Climpson had wondered greatly at the folly and wickedness of mankind.[18]

The man from the Psychical Research Society is probably based on Frederick Bligh Bond, whom Sayers had met in 1917. Bligh Bond was an archaeologist who excavated the ruins of Glastonbury Abbey after the spirit of a dead monk had supposedly told him where to dig. However, his firm belief in spiritualism had not prevented him from exposing several fraudulent mediums. In a letter to her parents, Sayers says he told 'hair-raising tales of how he and the head of the Psychical

Research Society are sent for to haunted houses, to Sherlock Holmes about for the haunters!'[19] Bligh Bond's account of his spirit guidance in excavating the Glastonbury ruins was published by Sayers' employers, Blackwell's, in the following year. Following this, the unfortunate Bligh Bond was promptly dismissed from his job and barred from the Glastonbury site by the Church of England, which did not want to be seen to be associated with spiritualism.[20]

Sayers uses Miss Climpson's situation to set up a moral dilemma: Miss Climpson is a devout Christian, and feels thoroughly guilty about producing fraudulent spiritualist phenomena, both because of the deception involved and because of the connection with the occult. She soothes her conscience by telling herself that it is for a worthy end: her efforts will save Harriet Vane from being wrongfully convicted of murder. Her deception presents an interesting situation – in theory a fake medium is someone who would be likely to be unmasked by a detective, rather than being a detective herself. After the will is found, Miss Climpson reverts to a more conventional rôle. She feels it is her duty to put her client on her guard against another fake medium, whom she describes to Lord Peter as being, 'as great a charlatan as I AM!!! and without my *altruistic* motives!!!'[21]

Agatha Christie uses a similar device in *Peril at End House* (1932), where Poirot arranges for all the murder suspects to attend a seance. Poirot's friend Hastings acts as a remarkably unconvincing medium. Poirot arranges for the supposed murder victim (who is in fact still very much alive) to appear at the seance in suitably ghostly fashion. This leads to some very interesting confessions and the revelation of the murderer's identity, as well as the arrest of a forger and the unmasking of a drug dealer.

Christie's *The Sittaford Mystery* (1931) begins at a seance where the spirit of a Captain Trevelyan informs the sitters that he has just been murdered. It turns out that Captain Trevelyan has indeed been murdered at or near the very time the spirit message came through. Throughout the book two mysteries run in parallel: who committed the murder and what is the explanation of the spirit message? Various explanations are suggested, ranging from clairvoyance and auto-suggestion to telepathy. One character even suggests asking Sir Arthur Conan Doyle's opinion. In fact the murderer created the spirit message by purely natural means in order to establish an alibi, and a perfectly logical solution to the 'supernatural' mystery has been obscured by a shoal of red herrings. Like the novel's characters, the reader is cleverly led to believe the information given at the seance, which is just what the murderer wants.

In both of these novels Christie's use of spiritualism is perfectly straightforward. The seances are definite fakes. She used the seance situation more ambiguously in some of her short stories. *The Hound of Death*, which was also written in the 1930s, is a collection of short stories about the supernatural. The stories are an odd mixture – some treat the supernatural as a fact, but one shows how easily people who believe in it can be exploited by fraudulent mediums. 'The Blue Jar' features a gullible young man who is tricked by a fraudulent medium. This story gains most of its impact from its context. The supernatural aspects appear convincing to the reader because the story is in a collection of supernatural tales: in a collection of detective stories the deception would be obvious. This authorial sleight-of-hand reflects the techniques used by fake mediums themselves. As Richard Wiseman's recreations of seances at the University of Hertfordshire have shown, even the most cynical of people have their scepticism dulled by the darkness and atmosphere of expectation which precedes any 'supernatural manifestations' in a seance.

In Christie's 'The Last Seance' a genuine medium is destroyed by her own powers. She makes the spirit of a dead child materialise and when the sitter touches the spirit, the medium dies. This story reflects many mediums' insistence that no-one should touch the 'spirits' without their consent, as this would supposedly result in death or serious injury to the medium. In practice, people who broke this rule often found themselves clutching remarkably solid 'spirits' who bore a distinct resemblance to the medium. In December 1873, for example, a Mr Volckman disrupted a seance by grabbing hold of a 'spirit' calling herself Katie King, who turned out to be none other than the medium Florence Cook, supposedly bound in a locked cabinet.[22] At another seance a sitter grabbed hold of a very substantial 'spirit' who proved to be the medium Miss Woods on her knees partly undressed and draped in muslin.[23] In 'The Last Seance', one of the characters explains the solid nature of such 'spirits' by arguing that,

> a spirit, to manifest itself, has to use the actual physical substance of the medium. You have seen the vapour or fluid issuing from the lips of the medium. This finally condenses and is built up into the physical semblance of the spirit's dead body.'[24]

This is an admirably lucid and concise explanation of a well-known spiritualist belief. It is possible that Christie had in mind the American medium Mina Crandon, known professionally as 'Margery' whose exploits received a great deal of publicity in the mid-1920s, when a team of investigators including Houdini and Conan Doyle tried to establish whether or not she was genuine. Ruth Brandon's book on spiritualism includes some truly repulsive photos of Margery in action. In one of them, an extra hand, supposedly made out of ectoplasm, appears to be emerging from her stomach.[25]

In another Christie short story, 'The Red Signal' a medium warns three clients 'Don't go home tonight'. The warning comes true for all three. The first man returns home and is murdered, the second returns home to find he has been framed for the murder and the third (who actually committed the murder) returns home and is arrested. One of the characters also has a premonition of danger immediately before the seance. The seance itself is described in distinctly sceptical terms. During the first part of the seance, 'messages were given from vaguely described relatives, the description being so loosely worded as to fit almost any contingency'[26] – a well known technique of fake mediums, and one used by Miss Climpson in *Strong Poison*. The medium herself is described in unpleasantly snobbish terms which reflect the views of the distinctly upper-class sitters: she is 'a plump middle-aged woman, atrociously dressed in magenta velvet, with a loud rather common voice', who believes that her own predictions are 'nonsense'.[27]

Christie also created a supernatural detective, Mr Harley Quin [*sic*], who appeared in a collection of short stories in 1930. Mr Quin is a supernatural being – supposedly an incarnation of Harlequin – who acts via a human intermediary, Mr Satterthwaite. Seances feature in two of the stories. In 'The Bird with the Broken Wing' a group of young people take part in a table-turning session as a joke, but a very real spirit message comes through for Mr Satterthwaite. In 'The Voice in the Dark' a dead woman's voice is heard at a seance. There is a natural explanation for this: the woman is actually still alive but going under a different name, and she is one of the sitters at the seance. As this explanation involves the woman in question having a split personality and a 50-year spell of amnesia, the 'natural' explanation is not much more convincing than the supernatural one.

F. Tennyson Jesse's 1931 detective Solange Fontaine also has dealings with fraudulent mediums. She attends a seance, where

> no less a person than St Elizabeth of Hungary appeared to us and wrote us messages in English on a little tear-off pad. Unfortunately, when grappled with by two men, friends of my own, who were present, St Elizabeth turned out to be a man – the medium in fact, dressed in white muslin.[28]

This may have been based on fact. Ruth Brandon reproduces a 1907 recipe to make a 'spirit veil' as used by a fraudulent male medium when impersonating female spirits. This involved several yards of fine silk soaked in benzine, lavender oil and luminous paint. The man who gave the recipe claimed to have used it when impersonating 'Cleopatra and other queens'.[29] Fontaine is later aided in her investigations by a self-confessed fraudulent medium who, to her own amazement, solves a murder case when she channels a perfectly genuine message from the supposed victim.

Mediums also appear in more recent detective fiction. In Paul Gallico's *The Hand of Mary Constable* (1964),[30] the detective (who, appropriately, is called Alexander Hero) is described as the 'chief investigator of the British Society for Psychical Research [and] an independent private detective of the occult ... an occupation which called for a thorough grounding in normal and abnormal psychology, physics, chemistry, biology, photography, magic, sleight-of-hand [and] laboratory procedure.'[31] Even Sherlock Holmes could not boast such a catalogue of accomplishments! Hero exposes a fraudulent medium who has produced the impression of the supposed 'spirit hand' of a dead child in wax, complete with correct fingerprints. Less sophisticated versions of this trick had been a favourite with earlier mediums. After Conan Doyle died in 1930, one medium, Valiantine, had produced Conan Doyle's 'spirit thumbprint' for his widow. When checked, the print turned out to be not Conan Doyle's thumb but Valiantine's own big toe.[32] The methods used by Gallico's medium are similar to those described by Dorothy L. Sayers, in *Strong Poison*, but with all the advantages of modern technology. For example, she uses infra-red light to move around freely in the darkness of the seance room.[33]

Gallico's book emphasizes that it is men of scientific training and intelligence who are fooled by the medium, because her simple sleight-of-hand tricks are outside their experience. A similar point was made by J.N. Maskelyne, a Victorian magician and investigator of psychic phenomena, who commented that, 'no class of men can be so readily deceived by simple trickery as scientists. Try as they may, they cannot bring their minds down to the level of the subject.'[34] As modern investigators such as James Randi have pointed out, their obsession with looking for a scientific explanation for paranormal phenomena can blind scientists to the use of simple conjuring tricks to produce supposedly supernatural effects.[35]

Although Gallico's medium is exposed as a fraud, the book's attitude towards spiritualism is distinctly ambivalent. Hero is described as being 'as eager for genuine proof of life in the hereafter as he was active in destroying the charlatans of spiritualism who preyed upon the bereaved and ignorant.'[36] He is keen to emphasize that his exposure of one fraudulent medium does not disprove spiritualism in general.[37] By the 1960s, spiritualism was not such a topical or controversial issue as it had been in the 1930s, so it is easier for the detective to be tolerant. This atmosphere of tolerance becomes even more marked in more recent fiction dealing with spiritualism. The growth of New Age beliefs has led to an interest in the occult, and spiritualism has become a 'respectable' subject for mainstream fiction such as Victoria Glendinning's novel *Electricity* (1995)[38] and A.S. Byatt's novella 'The Conjugial Angel' in *Angels and Insects*.[39] This is reflected by a more sympathetic portrait of mediums in some British detective fiction.

In *A Killing Kindness* (1980), Reginald Hill reverses the usual convention of not revealing the murderer's identity until the last chapter. The book's very first sentence names the murderer, but this is done in such a way that the reader does not realise it. This sentence is spoken by a medium in a trance. Although she is convinced that it is the voice of the murder victim speaking through her, even the medium herself does not realise the significance of what she is saying, nor do her audience. Transcripts of the seance are given to various characters in the novel, but no-one apart from the murderer realises the true significance of the medium's words.[40]

Hill's medium is a cultured Romany woman with a great knowledge of supernatural lore, unlike the badly-educated, vulgar charlatans of 1930s fiction. Although this medium is portrayed as perfectly genuine, the way Hill sets up the situation could be seen as the authorial equivalent of the sleight-of-hand tricks used by fake mediums. In the best tradition of detective stories, the author uses his professional skills to play tricks on the reader. Hill sets up a tension between character and narrative technique. His medium is portrayed as being open and honest, but the techniques he uses to portray her are deliberately misleading and full of trickery in order to prevent the reader from realising that she has in fact named the murderer.

In a sense, the author functions as the medium's agent or manager, working behind the scenes to stage her tricks. In *Revelations of a Spirit Medium*, published in 1891, the author, who claims to be a former fraudulent medium, warns,

> Do not forget the 'manager' in your search. He or she is never searched, or never has been, up to date, which has been the cause of

> many a failure to find the 'properties' of the 'medium' when the 'seance' was given in a room and 'cabinet' furnished by a stranger and skeptic.[41]

In writing about mediums, a skilled author can manipulate the reader in the same way that the manager would manipulate the medium's audience, playing on their preconceptions and expectations.

One of John Mortimer's Rumpole short stories, 'Rumpole and the Soothsayer',[42] deals with the exposure of a fake medium. The 'spirit voices' produced at his seances are in fact all produced by a confederate speaking through an intercom from another room. However, there is a twist at the end, when the medium discovers that the intercom was broken, so that at least some of the messages he was sure were fake are in fact genuine. This reverses the convention of detectives exposing mediums as frauds: in this story a medium who he thinks is fraudulent is proved to be genuine. Ironically, he foretells his own death but does not believe his own prediction – a modern twist on the Cassandra story?

In Carol O'Connell's *Mallory's Oracle* (1994), a group of elderly women who attend a weekly seance are murdered one-by-one. The medium, Redwing, is an impressive but unsympathetic figure who has been arrested three times for extortion and fraud, and is suspected of a range of crimes from child abuse to insider dealing. Her criminal career culminates in a murderous attack on the detective, Kathy Mallory. Unlike the mildly comical mediums of much 1930s fiction, Redwing is a sinister and imposing figure who exerts an almost hypnotic power over her victims. However, her table-turning performance bears a marked similarity to that of Miss Climpson. At a seance, Mallory notices that,

> When [Redwing] put her hands on the table, I saw the rings digging into her fingers. Then I saw the two ripples in the tablecloth where her rings had unhooked the pins under the material. All she had to do was lift.[43]

Mallory's cynicism is temporarily shaken during the seance when Redwing is supposedly possessed by the spirit of Mallory's recently murdered adoptive father, but Mallory later convinces herself it was trickery. However, a lingering unease about Redwing and other supposed psychics is felt throughout the book: they are seen as dangerous figures who inspire fear rather than ridicule.

A more benevolent mediumistic figure features in the sequel to *Mallory's Oracle, The Man Who Lied to Women* (1995). Mallory and her associate Charles Butler are heavily influenced by their memories of Malakhai, a magician who took up 'debunking paranormal frauds'[44] after he retired from the stage. Before retirement, Malakhai perfected an illusion which would be the envy of any fake medium: the spirit of his dead wife supposedly aids him with his stage act: 'after the audience got comfortable with the idea that she was not only invisible but dead, things began to float through the air as she handed him one thing and another.'[45] However, for all his scepticism, Malakhai himself believes in the very illusion he has created, and 'only created the flying-object illusion in the act so the audience could see her too.'[46] Charles comes to believe in a similar manifestation: he attempts to recreate the spirit of the murder victim in the hope that she will help him solve the case. He feels he has been successful, and the reader is left in some doubt as to how far this is an illusion. In combining the roles of detective and medium, has Charles literally made the dead speak?

The relationship between spiritualism and detective fiction has been a changing and uneasy one. The medium varies from being a figure of fun to a figure of supernatural power, and from being an undoubted fraud to being a genuine psychic whose power to make the dead speak is greater than that of the detective. In most of the examples I have given, the detectives are male and the mediums female, so the medium can also be seen as a threat to male power. In the Comtean tradition, women's mental powers were linked with instinct, spirituality and the supernatural, whereas men's mental powers were linked with science and logic. Spiritualism had been founded by women and was one of the first Western religions to place women in a central public rôle. Spiritualists believed that women were more sensitive to spirit influences than men. As Ann Braude points out, 'Spiritualism made the delicate constitution and nervous excitability commonly attributed to femininity a virtue and lauded it as a qualification for religious leadership.'[47] This unease with the female medium's rôle in a supposedly logical and scientific male-dominated society is summed up in *The Hand of Mary Constable*, whose author comments,

> The world in recent years had grown a good deal more sceptical about life in the hereafter, both in the religious as well as the occult sense. Yet underlying every man's complacency was still to be found that residue of terror that one day proof might establish survival of

some heretofore hidden forces more powerful than the overweening ego of man.[48]

In closing, I would like to comment that mixing spiritualism and detective fiction is not necessarily successful. Dickens' unfinished novel *Edwin Drood* – arguably one of the earliest detective stories – was completed after his death by an American medium who claimed that his spirit had dictated it to her. Dickens' son commented, 'I never saw this preposterous book, but I was told that it was a sad proof of how rapidly the faculties ... deteriorate after death!'[49]

Notes

1 Peter Lovesey, *A Case of Spirits* (repr. Harmondsworth: Penguin 1977), p. 30.
2 Arthur Conan Doyle, *The New Revelation* (London: Hodder & Stoughton, 1918), p. 19.
3 Ibid., p. 31.
4 Conan Doyle, *Memories and Adventures* (London: John Murray, 1930), p. 102.
5 Charles Higham, *The Adventures of Conan Doyle: The Life of the Creator of Sherlock Holmes* (New York: W.W. Norton, 1976), pp. 116–17.
6 See Higham, *The Adventures of Conan Doyle*, pp. 117–18 and Kelvin I. Jones, *Conan Doyle and the Spirits: The Spiritualist Career of Sir Arthur Conan Doyle* (Wellingborough: Aquarian, 1989), p. 80.
7 Conan Doyle, *Memories and Adventures*, p. 138.
8 Arthur Conan Doyle, 'The Story of the Brown Hand', *Strand Magazine* (May 1899), repr. in *The Supernatural Tales of Sir Arthur Conan Doyle* (London: W. Foulsham & Co., 1987).
9 Arthur Conan Doyle, 'Playing With Fire', *Strand Magazine* (March 1900), repr. in ibid..
10 Jones, *Conan Doyle and the Spirits*, p. 179.
11 Arthur Conan Doyle, *Our American Adventure* (London: Hodder & Stoughton, 1923), pp. 179–81.
12 Harry Houdini, *Houdini: A Magician Among the Spirits* (1924, repr. New York: Arno Press, 1972), pp. 150–8.
13 Conan Doyle, *Memories and Adventures*, p. 448.
14 Arthur Conan Doyle, 'The Sussex Vampire' in *The Case-Book of Sherlock Holmes*, repr. in *Sherlock Holmes: The Complete Short Stories* (London: John Murray, 1959), p. 1179.
15 Tom Gallon, 'The Spirit of Sarah Keech', in *The Girl Behind the Keys* (London: Hutchinson, 1903).
16 Lilian Wyles, *A Woman at Scotland Yard: Reflections on the Struggle and Achievements of Thirty Years in the Metropolitan Police* (London: Faber & Faber, 1952), p. 171.
17 Dorothy L. Sayers, *Strong Poison* (repr. Sevenoaks: Coronet, 1989), chapters 17–18, pp. 166–87.
18 Sayers, *Strong Poison*, p. 162.

19 Letter to her parents, 2 August 1917, repr. in *The Letters of Dorothy L Sayers, Vol. I, 1899–1936: The Making of a Detective Novelist*, ed. Barbara Reynolds (London: Hodder & Stoughton, 1995), p. 134 (see also p. 133).
20 The Visitor Centre at Glastonbury Abbey has reproductions of some of Bligh Bond's 'spirit drawings' on display. For other biographical information, I am indebted to Kathryn Denning's lecture on spiritualism and archaeology at the 'Victorian Supernatural' conference held at the University of North London in November 1995.
21 Sayers, *Strong Poison*, pp. 191–2.
22 Ruth Brandon, *The Spiritualists: The Passion for the Occult in the Nineteenth and Twentieth Centuries* (London: Weidenfeld & Nicolson, 1983), pp. 109–13.
23 *Spiritualist*, 28 September 1877, cited by Ronald Pearsall, *The Table Rappers* (London: Michael Joseph, 1972), p. 98.
24 Agatha Christie, *The Hound of Death* (repr. London: Fontana, 1986), p. 169.
25 Brandon, *The Spiritualists*, illustrations between pp. 212 and 213. The investigation of 'Margery' is described in chapter 6, 'Magicians Among the Spirits', pp. 175–89.
26 Christie, *The Hound of Death*, p. 30.
27 Ibid., pp. 29 and 31.
28 F. Tennyson Jesse, *The Solange Stories* (London: Heinemann, 1931), p. 10.
29 David Abbott, *Open Court*, March 1907, cited by Brandon, *The Spiritualists*, pp. 100–1.
30 Paul Gallico, *The Hand of Mary Constable* (London: Heinemann, 1964).
31 Ibid., p. 6.
32 Higham, *The Adventures of Conan Doyle*, pp. 337–8.
33 Gallico, *The Hand of Mary Constable*, p. 250.
34 *Daily Chronicle*, 29 October 1895, quoted by Brandon, *The Spiritualists*, p. 256.
35 See for example, James Randi, *Flim-Flam! Psychics, ESP, Unicorns and Other Delusions*, (New York: Prometheus Books, 1987) and *James Randi: Psychic Investigator* (London: Boxtree, 1991).
36 Gallico, *Hand of Mary Constable*, p. 6.
37 Ibid., pp. 270–1.
38 Victoria Glendinning, *Electricity* (London: Hutchinson, 1995).
39 A.S. Byatt: 'The Conjugial Angel', in *Angels and Insects* (London: Chatto & Windus, 1992).
40 Reginald Hill, *A Killing Kindness* (1980, repr. London: Grafton, 1987).
41 'A Medium', *Revelations of a Spirit Medium, or Spiritualistic Mysteries Exposed: A Detailed Explanation of the Methods Used by Fraudulent Mediums* (St Paul, Minnesota: Farrington & Co, 1891), repr. in facsimile eds Harry Price and Eric J. Dingwall (London and New York: Kegan Paul, 1922), quoted by Brandon, *The Spiritualists*, p. 268.
42 John Mortimer, 'Rumpole and the Soothsayer', in *Rumpole on Trial* (Harmondsworth: Penguin, 1992), pp. 152–62.
43 Carol O'Connell, *Mallory's Oracle* (1994, repr. London: Arrow, 1995), p. 164.
44 Carol O'Connell, *The Man Who Lied to Women* (1995, repr. London: Arrow, 1996), p. 72.
45 Ibid., p. 72.

46 Ibid., p. 73.
47 Ann Braude, *Radical Spirits: Spiritualism and Women's Rights in Nineteenth Century America* (Boston: Beacon Press, 1989), p. 83.
48 Gallico, *The Hand of Mary Constable*, p. 7.
49 Brandon, *The Spiritualists*, p. 56.

7

The Locus of Disruption: Serial Murder and Generic Conventions in Detective Fiction

David Schmid

Although various genres of popular fiction have incorporated the theme of serial murder successfully in recent years, more traditional forms of detective fiction have not done so. The incompatibility of serial murder and traditional detective fiction is not surprising, as serial murder has a number of characteristics that are antithetical to the conventions of the traditional detective fiction genre.[1]

The serial killer, who typically murders for non-rational motives and who usually murders strangers, challenges some of the most cherished assumptions of traditional detective fiction, namely the belief that murderers always have 'rational' motives for murder, such as greed or jealousy, and the belief that murderers always have some kind of personal relation with their victims. The presence of a serial killer thus problematises the solution of the central mystery of classical detective fiction, the mystery of the killer's identity. The marks of this problematisation are clearly visible in the narrative structure of detective fiction. Unwilling to ignore serial killers altogether, detective fiction writers have attempted, with varying degrees of success, to develop ways both of using serial murder and of distancing their detectives from a phenomenon that is so subversive of their genre's usual methods and assumptions.

I will demonstrate these points by discussing two texts. The first, Agatha Christie's *The ABC Murders* (1935), toys with the idea that such a thing as a truly motiveless serial killer might exist, but then ultimately rejects that idea in the interests of maintaining generic coherence. The second, P.D. James's *Devices and Desires* (1989), also attempts to integrate a serial killer into its narrative structure, but again has to dispense with that killer when his presence becomes too subversive of the formal conventions of the genre. The simultaneous appeal and difficulty of representing serial murder in classical detective fiction reveals a number of interesting points about the formal constraints of the genre.

The ambiguous appeal of serial murder to detective fiction writers lies in the recognition that seriality is both fundamental to and disruptive of the founding ideology of the genre, that is, the search for rationality and order in a world disrupted by criminal violence.[2] The detective's victory over the irrationality of murder, his/her success in re-establishing a sense of ontological and ideological stability is only ever temporary because 'yet another murderer, yet another bundle of contradictory clues, lurks around the corner'.[3] Even if the Great Detective manages to solve an individual murder and thus to restore the image of a benevolent and knowable world, this same image will be disrupted once more almost instantly.

In as much as the structure of murder can be thought of as a continuous and potentially infinite series, serial murder is therefore the 'irrational' of detective fiction, the element which persistently disrupts the drive toward order in the genre by ensuring that 'mystery always returns'.[4] Given the disruptive nature of serial murder, one might expect readers and writers of detective fiction to view it negatively. Instead, they view such cases as a challenge, something to relieve *ennui*. Serial murder exemplifies the 'perfect murder' that David Lehman describes:

> the perfect murder is not the one that goes unsolved but the one that most strenuously taxes the detective's imagination ... it calls for an intricate design, for huge reserves of guile under pressure, and it makes the successful sleuth seem like a miracle worker. What's more, it has usually been embellished in some fanciful way; the perpetrator has prepared it with all the trimmings of a labour of love. The perfect murder is, in short, a work of art, diabolical in its ends and entertaining in its means – a riddle and a cipher.[5]

This double attitude towards serial murder, seen as both an awesome challenge and an object of aesthetic contemplation, can be found in Agatha Christie's *The ABC Murders*, an example of the type of detective fiction that tends to treat serial murder primarily as a way to 'enhance the surprise of the denouement'.[6] In other words, Christie, although intrigued by the possibility of serial murder, ultimately does not take this possibility seriously because she *cannot*. Christie's fictional world is predicated not only on the inevitability, but on the *necessity* of the restoration of order, and although serial murder may be experimented with as an interesting narrative twist, it cannot be allowed to realize its potential to destabilise radically the rationalism of the traditional detective fiction genre.

The plot of *The ABC Murders* concerns a series of apparently random and unmotivated murders committed by a murderer known only as 'ABC'. The murderer gives prior notice of all his murders in the form of taunting letters sent to Christie's hero, Hercule Poirot. The murders are distinguished principally by being committed in an alphabetical sequence that encompasses the first and last names of the victims and the town in which they live. The first is Ann Ascher of Andover, the second Betty Barnard of Bexhill-on-Sea, the third Sir Carmichael Clarke of Churston, and so on. The murderer leaves a copy of the *ABC Railway Guide* at each murder site.

That such a series might be regarded as an object of aesthetic interest and even pleasure is suggested by a conversation between Poirot and his Watsonesque sidekick Captain Hastings after the arrival of the first letter from ABC. The conversation concerns what kind of murder each would choose to investigate if one were able to order a murder as one orders a fine meal. Their exchange is a good example of the self-consciousness about the conventions of detective fiction that one finds throughout Christie's work, as we see when Poirot describes Hastings's choice as a 'résumé of nearly all the detective stories that have ever been written'.[7] Poirot's own choice is what he describes an *intime* crime: 'A very simple crime. A crime with no complications. A crime of quiet domestic life ... very unimpassioned.' (p. 23) After Hastings complains that this does not sound very exciting, Poirot accuses him of having a 'melodramatic soul' and of wanting 'not one murder, but a series of murders' (p. 24). Hastings admits to this preference, explaining that 'a second murder in a book often cheers things up.'

In spite of Poirot's preference, Christie gives her readers what Hastings wants instead. This is perhaps because Christie realizes, as Hastings seems to, the 'entertainment value' of a series of murders. This view of serial murder as aesthetically entertaining or enthralling can be found throughout the text. For example, when Hastings first hears of the murder of Ascher, he is disappointed because it does not meet his expectations: 'I had expected something fantastic – out of the way! The murder of an old woman who had kept a little tobacco shop seemed, somehow, sordid and uninteresting.' (p. 25) Yet with the discovery of the *ABC Railway Guide* near Ascher's body, Hastings says that the 'sordid crime took on a new aspect' (p. 32), that is, it became aesthetically interesting by virtue of being *outré*.

The aestheticised appreciation of the murders goes hand-in-hand with an awareness of how the murders challenge Poirot (and this may of course be another reason why Christie chose this plot). Poirot initially interprets the letters as a personal challenge from ABC, whom he assumes to be someone he has wronged in the past. For much of the book, Poirot does not seem equal to the challenge. He is denied the chance to investigate the murders in his customarily proactive way because the apparently unmotivated nature of the murders deprives him of clues in the normal sense. Poirot must instead be frustratingly *re*active, waiting for the ABC murderer to strike again after receiving his latest letter.

Poirot's helplessness is satirically hinted at in an exchange with Hastings as they are leaving Andover after having investigated Ascher's death. In answer to Hastings's inquiry, 'Well?', Poirot answers, 'the crime... was committed by a man of medium height with red hair and a cast in his left eye. He limps slightly on the right foot and has a mole just below the shoulder-blade.' (p. 54) Typically, it takes Hastings a moment to realize that Poirot is joking. But in spite of the joke, Poirot has to admit with uncharacteristic frankness, 'I do not know what the murderer looks like, nor where he lives, nor how to set hands upon him.' Such a profession of ignorance might seem to undermine Poirot's status as the representative of rationality, but his statement is stabilised by the reader's knowledge of the generic conventions of this type of detective fiction, conventions which dictate that Poirot will eventually arrive at a (re)solution.

Because of the nature of the crimes, Poirot realizes early on that he must rely not on the 'leads' traditional to a murder investigation, clues such as the 'match end, the cigarette ash, the nailed boots' (p. 56), on which Poirot says Hastings relies excessively. Instead, he must look to the psychological make-up of the murderer as revealed by his crimes.[8] This concentration on the psychology of the murderer means that *The ABC Murders*, rather than prioritising the chain of evidence that leads to the killer, emphasizes instead how one is to conceptualize the possibility of a 'mad' and unmotivated random serial murderer. The motif of madness is introduced at an early stage of the book when Poirot shows Hastings the first letter from ABC. When Hastings reads the letter, his response is immediate and predictable: 'Some madman or other, I suppose.' (p. 16) Similarly, when ABC's second letter arrives, Hastings exclaims, 'we're up against a homicidal maniac!' (p. 59) This phrase is subsequently used throughout the book by a number of characters, and it is clear that the 'ordinary person' is meant to make an automatic, commonsense equation of this type of random murder and madness.

The first substantive discussion of madness comes after the murder of the first victim and immediately before the murder of the second. The investigation, such as it is, is completely stymied. The police and Poirot have got together for a council-of-war and Poirot suggests that the murderer might be choosing his victims according to an alphabetical pattern. At this point Dr Thompson, whom Christie introduces as a 'famous alienist' (p. 60), reminds us that 'we're dealing... with a madman [who] hasn't given us any clue as to motive' (p. 61). The police superintendent wonders aloud whether the mad can be said to have motives, and Dr Thompson replies with asperity: 'Of course he has, man. A deadly logic is one of the special characteristics of acute mania.'

This discussion not only reinforces the connection between this type of murder and madness, it also establishes the centrality of the question of motivation. This is a subject that Poirot will return to again and again, arguing, 'if we knew the exact reason – fantastic, perhaps, to us – but logical to him – of *why* our madman commits these crimes, we should know, perhaps who the next victim is likely to be.' (p. 93) The way in which the question of motivation is posed, however, is determined by the assumptions Poirot and the other characters make about the 'chain' or 'series' type of murderer they are dealing with.

These assumptions are presented in another discussion between Poirot and the police after the second murder. By now they accept the existence of the 'alphabetical complex', but what troubles Poirot is not the validity of ascribing a pattern to murder *per se*, but whether the pattern in the current case is consistent. The first assumption that governs the conversation between the investigators is that even apparently unmotivated or irrational murderers kill only for certain reasons, either to remove people who are a source of annoyance or obstruction to them, or because the murderers think that they have some sort of mission to kill a particular class of victim. Based on this assumption, what troubles Poirot is that 'if the victims are alphabetically selected, then they are not being removed because they are a source of annoyance to him personally' (p. 92), because, as Poirot says, it would be too great a coincidence for the victims both to die in alphabetical sequence and to be personal acquaintances of the murderer.

So far, Poirot's logic appears to be sound, but what he excludes from consideration in this model of irrational murder is the possibility that murderers *do* exist who choose their victims in a random fashion. Indeed, Dr Thompson explicitly states that 'there isn't such a thing as a murderer who commits a crime at random' (p. 92). Once this claim is established, it allows a second claim to be made: if there is no such thing as a truly random murderer, the ABC murderer must have another principle of victim selection, 'apart from the merely alphabetical one' (p. 93). Simply by discussing irrational murder and its motives at all, Christie is unusual among detective fiction writers of the time. According to Julian Symons, in traditional detective fiction,

> it was accepted that the motives for all crimes should be personal, and within that context rational. They should not be committed ... by somebody merely insane. It was permissible that the people in a story should *think* that a crime was irrational ... but the reader knew that there would always turn out to be a personal motive.[9]

In playing with the idea of an irrational murder, but then anchoring her story back into the rational Christie shows her combination of experimentation and tradition.

In order to determine what the other, non-alphabetical principle might be, Poirot concentrates on what he sees as the inconsistencies of the case, both in the pattern that ABC apparently uses, and in the psychological make-up of the murderer. Poirot's view of these inconsistencies is summed up by his description of ABC as 'the magnanimous murderer' (p. 93). By this he means that there is an apparent inconsistency between the murderer's great ruthlessness and cruelty in killing complete strangers and the letters that announce his crimes, which – Poirot suspects deliberately – divert suspicion from members of the victims' families, who would otherwise be prime suspects.

Poirot's problems with the inconsistencies in the case ironically come to a head at the moment the case is apparently solved, with the arrest of Alexander Bonaparte Cust. There seems to be a cast-iron case against Cust. Apart from the obvious suggestiveness of his initials, the murder weapon used in the fifth and last murder is found at his lodgings, along with piles of *ABC Railway Guides* and the typewriter that is proved to have been used to write the letters. Despite the overwhelming evidence against Cust, Poirot remains unconvinced that he is the murderer because of the psychological inconsistencies that he has been emphasizing throughout the investigation. As he says of Cust,

> We know that he was quiet and retiring – the sort of man that nobody notices. We know that he invented and carried out an intensely clever scheme of systematized murder. We know that he

> made certain incredibly stupid blunders. We know that he killed without pity and quite ruthlessly. We know, too, that he was kindly enough not to let blame rest on any other person for the crimes he committed. If he wanted to kill unmolested – how easy to let other persons suffer for his crimes. Do you not see, Hastings, that the man is a mass of contradictions? Stupid and cunning, ruthless and magnanimous – *and that there must be some dominating factor that reconciles his two natures.* (p. 199)

A typical Agatha Christie novel attempts to leave the reader with the 'gift of a tidy world, a closed book in which all questions have been answered'.[10] Poirot tries to bring about this type of closure, and to resolve the contradictions in Cust's nature, by way of the most time-honoured element in the genre, the explanation scene, in which Poirot reviews all the main themes and facts of the case and reveals the identity of the murderer. Poirot begins by restating his claim that the key issue in this case is not so much the murders themselves but the mind and motivations of the murderer. He goes on to explain how, given this emphasis, the alphabetical pattern was comparatively irrelevant next to the question of 'why did A.B.C. *need* to commit these murders?' (p. 217). This is where the inconsistencies trouble Poirot, in that they inhibit his seeing clearly into the mind of the murder.

The inconsistencies between Cust's character and the necessarily ruthless and calculating nature of the murderer lead Poirot to conclude that someone else must have committed the murders and set Cust up with the aid of the letters. In concentrating on the function of the letters, Poirot concludes that they had a dual purpose: to draw attention both to the suspect the murderer was setting up (that is, Cust) and to a *series* of murders rather than a single crime. As Poirot says, 'you notice an individual murder least… [when] it is one of *a series of related murders*' (p. 233). This line of reasoning leads Poirot to the following conclusion:

> I had to deal with an intensely clever, resourceful murderer – reckless, daring and an thorough gambler. *Not* Mr Cust! He could never have committed these murders! No, I had to deal with a very different stamp of man – a man with a boyish temperament (witness the schoolboy-type letters and the railway guide), an attractive man to women, and a man with a ruthless disregard for human life, a man who was necessarily a prominent person in *one* of the crimes! (p. 223)

At this point Poirot finally produces his murderer, Franklin Clarke, the brother of the third victim, Sir Carmichael Clarke. Franklin wanted to murder his brother because he was afraid that he would be cut out of the will and that all his brother's money would go to his assistant, Thora Grey. Franklin Clarke hit upon the idea of the alphabetical scheme when he met Cust by accident and realized that the shell-shocked war veteran would make a perfect fall guy.

If the line of reasoning that leads Poirot to Franklin Clarke seems somewhat strained, this reflects how far Poirot has departed from his normal method of accumulating physical evidence. Instead of relying on what we normally think of as evidence, Poirot's conclusion stands or falls by the psychological deduction he makes, aiming to convince the reader that Clarke is indeed the murderer because his psychological characteristics fit those that Poirot attributes to the ABC murderer. This method of deduction, however, is inevitably less concrete than that based on material evidence, and it requires more of an act of faith on the part of the reader than would normally be the case in a Christie mystery.

Franklin Clarke personifies what Roger Caillois has described as the way 'madness can be used to mask shrewdness ... an extremely sly murderer may find some advantage in suggesting that his crimes were the work of a maniac.'[11] The advantage to the writer of this type of presentation of madness in detective fiction, according to Caillois, is that 'the author gains all the fantasy and strangeness that madness brings with it without in any way giving up the privilege of reason when it comes time for a final explanation.' There could not be a better description of how Christie uses the possibility of a serial murderer killing according to an alphabetical pattern for most of *The ABC Murders*, and then stabilizes the 'fantasy and strangeness' of the pattern by making Franklin Clarke the murderer. Christie controls the anarchy and chaos that one might think would go with representing madness by emphasizing what Caillois calls 'the mechanical aspect of madness', which gives the murderer's actions 'a systematic character, the appearance of an absolutely imperious and externally determined necessity.'[12] Franklin Clarke's carefully designed and executed homicidal plan is meant to stand as the polar opposite of the random, 'unmotivated' attacks of a serial killer.

Christie's desire to expunge this randomness from her text explains the insistent tone of the pronouncements throughout *The ABC Murders* to the effect that a random serial murderer is an impossibility, but it is unclear whether Christie succeeds in controlling the spectre she has created in the interstices of her text. At one point in the investigation Poirot argues that 'it was not *chance* that provided A.B.C. with a victim in Betty Barnard. There must have been deliberate selection on his part – and therefore premeditation' (pp. 124–5). Of course, when Franklin Clarke is revealed as the murderer, we see that Poirot is right, but this incident is suggestive of the way in which Poirot must raise the possibility of a murderer who kills according to chance in order to discount it. Christie hopes that readers will be convinced by that discounting, but the 'residue of horror'[13] that results from the suggestion remains to make any sense of closure in Christie's text partial and unfulfilling.

If the reader remains unconvinced by Christie's attempt to integrate a serial killer into a genre whose narrative structure cannot assimilate such a figure, however, her solution actually seems extremely adroit when compared to P.D. James's handling of the same problem in *Devices and Desires*. Just as the ABC murderer challenges Poirot, so 'the Whistler' challenges James's detective, Adam Dalgliesh. Rather than bring her detective into direct contact with the serial killer (as Christie does), James instead chooses to resolve the tensions raised by the presence of the Whistler and Dalgliesh in a much more arbitrary and unsatisfying manner. James shows even more clearly than Christie the tensions that the serial killer generates within the narrative structure of traditional detective fiction.

The source of tension in James's text is the serial murderer she creates, the Whistler, who dominates the first half of her book. At the half-way point, the Whistler suddenly commits suicide, and the remainder of the book is devoted to the investigation of the murder of Hilary Robarts (staged by the murderer to make it look like the work of the Whistler). The Whistler's suicide is expedient rather than convincing; his demise has the air of a *deux ex machina* rather than having been brought about by any narrative necessity. In dispensing with the Whistler, James fights a rearguard action in defence of the generic conventions of traditional detective fiction, conventions that the serial killer seriously destabilises.

James's dismissal of the Whistler after he has served his purpose (that is, after he has 'introduced' the murder of Hilary Robarts) and her preference for Alice Mair over the Whistler as a murderer both derive from her persistent devotion to the traditional conventions of detective fiction, a devotion that James herself has commented on:

> The old conventions may still be retained. There will be a violent death; a limited circle of suspects all with motives, means and opportunity; false clues; and a tenable ending with a solution to the mystery which both author and reader hope will be a satisfying consummation of suspense and excitement.[14]

Devices and Desires displays many of the characteristics that James describes here. The violent death is that of Hilary Robarts and the circle of suspects is severely limited, restricted almost entirely to those present at a dinner given by Alice Mair and her brother Alex, who were the only people to hear enough about the Whistler's methods to be able to stage Robarts's murder convincingly. The Hilary Robarts investigation takes up only the second half of the book, however, and to get to this point, James attempts what is a significant departure for her, the representation of a serial murderer. The reason that representing a serial murderer is such a departure for James is that typically 'her criminals are not hard-core, underworld types but ordinary people pushed to extraordinary means to solve what they perceive to be problems. Their crime is generally a one-time occurrence; it is usually true that even the murderers among them would not be likely to repeat.'[15] Thus, the typical James villain could hardly be more different from the serial murderer, who kills to assuage a need or desire or simply because he enjoys it (rather than killing in order to solve a problem), and who kills repeatedly until he is caught. If we examine how James first introduces the Whistler to the reader and then has him commit suicide at a 'convenient' point in the narrative, we will see why James 'prefers' the murder of Robarts to those committed by the Whistler, and this will in turn allow us to see how serial murder, despite James's efforts, remains to undermine the narrative structure of her text.

The Whistler has a prominent place in *Devices and Desires*, but James carefully regulates this prominence by controlling both what information is given to the reader about the Whistler and how that information is presented. A description of the Whistler's murder of Valerie Mitchell opens the book, and this incident places the Whistler squarely at the centre of the narrative. One must note, however, that Valerie Mitchell is the Whistler's fourth victim, rather than his first. This is significant because joining the Whistler in the midst of his series of murders enables James to present a considerable amount of information about him – such as the fact that he has a car, that he strikes at night, and what his *modus operandi* is – as common knowledge in the area where the murders are taking place. The presence of this body of already 'accepted' knowledge protects James from having to go too deeply into the Whistler's motivations and also discourages her readers from doing so. By presenting his work in progress, as it were, James avoids exploring the origins of the Whistler's murderous impulses, except, as we shall see, in the crudest possible terms.

Apart from presenting the Whistler in the middle of his series of murders, James controls the reader's access to information about him by limiting the amount of contact Dalgliesh has with the investigation into the Whistler's killings. Dalgliesh does not really become involved with the case until after the Whistler is dead, and the investigation focuses on the murder of Robarts. The reason why James isolates Dalgliesh from the Whistler case can be found in how she characterizes serial murder.

Throughout *Devices and Desires*, serial murder is seen as different from 'normal' murder, both more deviant and more difficult to investigate. Dalgliesh describes serial murder cases as 'invariably difficult, time-consuming and frustrating, depending as they often did more on good luck than on good detection'.[16] He later expands on these thoughts: 'of all murders, serial killings were the most frustrating, the most difficult and the chanciest to solve, the investigation carried on under the strain of vociferous public demand that the terrifying unknown devil be caught and exorcised forever' (p. 70). Michael Holquist has described the typical image of the detective as the personification of mind and rationality,[17] and this is a description that seems made for Dalgliesh. Over the course of James's career, Dalgliesh has become her most durable series character, distinguishing himself in his cases through a mixture of erudition and educated guesses. But there is no doubt that Dalgliesh is best when investigating a certain type of murderer, a type whose murders reveal recognizable, though complex, mental traits. The Whistler undoubtedly has a motive, but it is non-rational, and his victim selection is dictated by chance rather than acquaintance. Faced with this type of murderer, dependent on good luck rather than good detection, Dalgliesh would be helpless and James protects her detective from direct involvement.

James prefers to concentrate on a murder which will allow Dalgliesh to explore a tangled skein of motivations, all of which could be traced back to intricate but ultimately comprehensible fears, resentments and desires. These qualities are summarised in James's description of the 'old conventions' of detective fiction, and are exemplified by the murder of Hilary Robarts. Apart from the question of how well it matches the 'old conventions', the key to James's concentration on the Robarts murder is the investigating officer Terry Rickards's comment that 'we're going to be dealing with intelligent suspects' (p. 194). Clearly, James's attraction to the Robarts murder and the people suspected of it has a class dimension.

This class dimension exists in the form of a comparison between the murderers themselves. Alice Mair is a writer and editor of cookbooks with a very healthy income. She lives with her brother, Alex, the director of a nearby nuclear power station, in a large, tastefully decorated house. The Whistler, Neville Potter, whom Rickards describes as a 'scrawny little sod' (p. 187), lives in a small, dilapidated caravan on a caravan site and works as a general handyman. The class contrasts between the murderers could hardly be clearer. Moreover, James never gives the Whistler the opportunity to defend or explain his actions, and his demise is presented as undignified, even squalid. His body is found in the back bedroom of a common seaside hotel and, being clad only in a T-shirt and underpants, he is explicitly denied the kind of dignity given to Mair.

From the first time that Dalgliesh meets Alice Mair, she is presented as an utterly self-composed and very intelligent person, so much so that even Dalgliesh finds her formidable. It is this intelligence and authority that James finds so attractive in Mair, as it implies that she would only commit murder for certain reasons. Towards the end of the book, when Meg Dennison, Mair's close friend, accuses her of murdering Robarts, Mair defends her actions in lengthy and complicated terms that ridicule the moral framework implied by the terms 'good' and 'evil'. Mair gives the impression that she could never commit a crime of passion, but only a crime of intellect that she can calmly rationalise. Partly because of Mair's intellectuality, James lets her keep that thing most precious to her, her dignity. Instead of being punished by the judicial system, Mair dies by her own hand when she sets fire to her house and perishes in the ruins.

James thus has many reasons for 'preferring' the murder of Hilary Robarts by Alice Mair to the murders committed by the Whistler, but it is important to her that Mair must still be seen to be punished. With Mair's suicide, James seems to uphold what Siebenheller has described as her 'strong belief in retribution'.[18] That this belief should be expressed in such a muted and ambivalent fashion is consistent with what James sees as the complexities of the Robarts case, and in particular of the murderer. Just as Alice Mair is a complicated, multi-layered individual, so the questions of why she murdered, her degree of guilt and the manner of her 'punishment' are all similarly ambiguous and opaque. But although James thus 'succeeds', on her own terms, in resolving the complexities of Alice Mair in a suitably open-ended and ambivalent fashion, *Devices and Desires* also contains a false – because it is overly tidy and unambiguous – resolution. The death of the Whistler provides a stark contrast to the open-endedness of Mair's demise, but James does not linger on this death. Indeed, she makes every effort to bury it deep in the middle of her book, but it proves to be remarkably resilient.

The false resolution comes in the form of the Whistler's suicide, and when Rickards and Dalgliesh go to view the body, James attempts to tie up all the loose ends of this section of the plot with a vengeance. In order to 'explain' him, Rickards refers to the Whistler's childhood and especially to his mother, who is introduced in order to 'solve' the mystery of why the Whistler carved the letter 'L' into his victims: 'You should see the mother. She's a right bitch that one. And do you know what her name is? Lilian. *L* for "Lilian". That's something for the trick-cyclist to chew over. She made him what he was.' (p. 190) Rather than attempting to discover what it was about the Whistler that made him direct homicidal rage against women (by, for example, exploring his family history in the same depth as she does that of Alice Mair), James chooses to sidestep the issue of responsibility and instead directs the blame at the Whistler's mother, who is criticised for being that most terrible of things, the dominant, suffocating and emasculating maternal figure.

If Rickards thus gives an overly tidy explanation of the Whistler's motive, it is interesting to note that this tidiness appears as a characteristic of the Whistler himself. He very conveniently leaves all the evidence of his crimes, together with a suicide note, on top of a chest of drawers. The Whistler's obsessive tidiness is clearly emblematic of the way in which James wishes to apply narrative closure to his case, but I think that what strikes the reader is the artificiality rather than the authenticity of this scene. The sense is of an issue deferred, rather than of a question resolved. The reader cannot help but wonder what might have happened if the Whistler had continued to kill and Dalgliesh had become involved in the case. This is precisely the scenario James wants to avoid.

James is ultimately unable to resolve the presence of a serial murderer in her text satisfactorily because she is not willing to face up to the amount of contradiction and uncertainty that such a figure introduces into the traditional detective fiction genre. This is not to say that James is unaware of these issues, only that she simply decides not to tackle the questions of how to explain a murderer with a non-rational motivational system convincingly, and of how a traditional detective investigates a non-traditional murderer whose capture depends largely on chance. As a result, James's text remains marked by an interesting tension between the randomness of serial murders and the predictability of the generic conventions of detective fiction, a tension that can only be 'resolved' or 'explained' through a refusal to examine in any depth the special characteristics of serial murder.

Christie and James develop different responses to a shared problem, namely of how to represent a serial killer, a figure who in many respects is the total opposite of the type of murderer we are accustomed to find in traditional detective fiction. Rather than committing a finite number of murders of people he or she knows for clearly recognizable reasons, the serial killer murders a potentially infinite number of victims who are strangers for motives that are typically non-rational. Given the problems that the serial killer raises for Christie and James, one might wonder why they would bother to write about this type of killer at all. The answer, I think, can be found in the desire of both writers to experiment with a genre they find simultaneously comforting and constricting. The serial killer seems to offer an exciting alternative to an 'ordinary' murderer, but neither Christie nor James is prepared to adapt the detective fiction genre enough to accommodate such a figure. In particular, their unwillingness to displace the central source of mystery from traditional detective fiction, that is, the question of the identity of the killer, dooms them to wrestle with ultimately irresolvable narrative problems. Christie chooses to have Poirot directly confront and then deny the possibility of a non-rational serial killer, while James keeps her detective more clearly distanced from such a figure. By refusing to accept the serial killer on his/her own terms, both Christie and James fail to find a way to integrate the serial killer into their narratives convincingly.

Rather than being personal, however, this failure is generic. The example of a book such as Thomas Harris's *The Silence of the Lambs* suggests that once you remove the mystery of the killer's identity by revealing it at the beginning of the text, the result is a police procedural or a thriller rather than a traditional whodunit. Both *The ABC Murders* and *Devices and Desires* suggest that the serial killer must be excluded from the genre in order to produce a text recognizable as detective fiction.

Notes

1 The phrase 'traditional detective fiction' needs to be emphasized here. Although many recent examples of crime fiction, such as the Kay Scarpetta novels of Patricia Cornwell, regularly feature serial killers, they do so in such a way that these novels are more police procedurals than classic 'whodunits'.

Thus writers such as Cornwell avoid the problems that plague Christie and James. See my comments on Thomas Harris's *The Silence of the Lambs* below.

2 The existence of the series of books, with one or more recurring characters, is another way in which seriality is imbricated in the institutional structure of the genre. See the chapter in this volume by Martin Priestman, 'Sherlock's Children: the Birth of the Series'.

3 Ernest Mandel, *Delightful Murder: A Social History of the Crime Story* (London: Pluto, 1984), p. 28.

4 Ibid., p. 27.

5 David Lehman, *The Perfect Murder: A Study in Detection* (New York: Free Press, 1989), p. 39.

6 David Richter, 'Murder in Jest: Serial Killing in the Post-Modern Detective Story', *Journal of Narrative Technique*, 19.1 (1989), 106–14.

7 Agatha Christie, *The ABC Murders* (New York: Pocket Books, 1941), p. 23. Henceforth references for quotations will be indicated in parentheses in the text. All emphasis is in the original.

8 I use 'his' advisedly. At an early stage in the investigation the police surgeon, Dr Kerr, says, 'psychologically speaking, I shouldn't say this was a woman's crime'. At this point, Poirot 'nodded his head in eager agreement' (p. 33). Thus the possibility of a female suspect is discounted without further serious consideration.

9 Julian Symons, *Mortal Consequences: A History from the Detective Story to the Crime Novel* (New York: Harper & Row, 1972), p. 3.

10 David Grossvogel, *Mystery and its Fictions: From Oedipus to Agatha Christie* (Baltimore: Johns Hopkins, 1979), p. 52.

11 Roger Caillois, 'The Detective Novel as Game', in Glenn W. Most and William W. Stowe (eds), *The Poetics of Murder: Detective Fiction and Literary Theory* (New York: Harcourt Brace Jovanovich, 1983), p. 3.

12 Ibid., p. 7.

13 Ross Macdonald, 'The Writer as Detective Hero', in Robin Winks (ed.), *Detective Fiction: A Collection of Critical Essays* (Englewood Cliffs, N.J.: Prentice-Hall, 1980), p. 180.

14 Bruce Harkness, 'P.D. James', in Bernard Benstock (ed.), *Art in Crime Writing: Essays on Detective Fiction* (New York: St Martin's Press, 1983), p. 119.

15 Norma Siebenheller, *P.D. James* (New York: Friedrich Ungar, 1981), p. 117.

16 P.D. James, *Devices and Desires* (New York: Warner Books, 1989), p. 11. Henceforth references for quotations will be indicated in parentheses in the text.

17 Michael Holquist, 'Whodunit and Other Questions: Metaphysical Detective Stories in Postwar Fiction', in Most and Stowe (eds), *The Poetics of Murder*, pp. 157–7.

18 Siebenheller, *P.D. James*, p. 100.

8
The Detective as Clown: a Taxonomy

Audrey Laski

It might, on the face of it, seem an unpromising enterprise to propose a family resemblance between such diverse series detectives as Father Brown, Inspector Ghote, Albert Campion, Peter Duluth, Adam Dalgliesh, Deputy Superintendent Dalziel and Marcus Didius Falco. The word which may conjure up the right connective images is clown, provided that the reader will recall what a range it encompasses, from Marcel Marceau's frail Bip to the robust fellow with the slapstick and the pail of paint who is the children's circus notion of a clown; it may also be helpful to think of the distance between Charlie Chaplin, the little man with a bowler hat, stick and big boots, and Charles Chaplin, the melancholy protagonist of *Limelight*; or, again, that the term clown is applied in Shakespeare's work to such diverse persons as Costard, Autolycus and Feste.

I see such detectives, and, indeed, many other male characters in fiction, as cut from a number of templates – a term preferable to the pejorative 'stereotypes' – which give them a likeness to each other, a kinship which may be summed up under the general heading myth – in this case the myth of the Clown – because, within these likenesses, there are resonances which go back to our earliest mythic ideas. Examining them does not support the idea that detective heroes must be shallow stereotypes, any more than studying the forms of the sonnet encourages the reader to think that 'Upon Westminster Bridge' or 'Shall I compare thee to a summer's day?' are doggerel; rather it gives a tool for the exploration of surprising subtleties and complexities. Moreover, the myths in a culture's fiction say something about that culture's perceptions and dreams, and so repay study.

The idea of making a taxonomy of character types under these mythic groupings arose initially from the reading of Victorian fiction, where there seemed to be two myths in a state of creative tension for much of the century, dramatising on the one hand the confident rise and rise of the self-made man and on the other a lingering regret for the decline of the glamour of the land-owning classes. The latter was expressed through what seemed appropriately described as the myth of the Aristocrat, the concept of a kind of innate superiority, which might have nothing to do with social birth: it provided a set of templates for characters set apart from the rest of humanity by some kind of exceptional quality; this might be extraordinary guilt, as in Byron's Manfred (the Satanic), or goodness, as in Lord Warburton (the Perfect Aristocrat), or attractiveness, as in Rickie Carstone (the Charmer), or utopianism, as in Angel Clare (the Idealist), or self-centredness, as in Sir Willoughby Patterne (the Egoist), or cleverness, good or evil, as in Sherlock Holmes and Professor Moriarty (the Benign and Malign Intellects).

As the last reference suggests, the Benign Intellect is the template of the myth of the Aristocrat most likely to furnish a series detective; the Egoist, the Charmer and the Idealist can rarely be expected to fill any other rôle than those of victim or murderer, and the Byronic Satanic usually occurs in Mills and Boon romances rather than thrillers, though in Dorothy Dunnett's remarkable series of historical novels, the Lymond saga, her hero Francis Lymond presents all the characteristic features of the Satanic through a series of adventures each of which centres on a mystery of some kind. Rex Stout's Nero Wolfe – 'I have no talents. I have genius or nothing'[1] – set apart by his monstrous bulk as well as his detective genius, is the supreme twentieth-century example of the Benign Intellect template in use, while Lord Peter Wimsey, in the later novels and particularly in *Gaudy Night* (1935), where his multitude of virtues almost sinks the vessel, illustrates the detective as Perfect Aristocrat.

The myth which offered an opposing response to the world in the Victorian novel was the myth of the Hero, the man who, unlike the Aristocrat, felt himself to be wholly at one with his fellow men, achieving nothing that they, with equal energy and determination, could not also achieve and retaining a vigorous independence of mind. This myth looks back to the heroic age of the sagas, and is the myth of maturity, as the myth of the Aristocrat looks back to the age of chivalry, had a powerful renaissance with the Romantic movement and is the myth of adolescence.

Unlike the myth of the Aristocrat, the myth of the Hero does not yield a number of diverse templates; the kinship between all these energetic, forward-looking men is very strong and obvious, as demonstrated by the work of Dick Francis. For though he only rarely, and always disappointingly, carries through a central character from one novel to another and these characters usually become detectives only through the pressure of circumstance, not by profession, Dick Francis constantly reinstantiates the Hero template in his remarkably popular thrillers.

It may be instructive to look more closely at one novel, *Flying Finish* (1966) where, unusually, his narrator–protagonist demonstrates Aristocratic rather than Heroic characteristics. The mythic Hero normally prefers fists to weapons and resists the temptations of 'wild justice', choosing to leave his revenge to society, and the majority of Francis's heroes follow this line, sometimes in the face of great provocation. Henry Grey strikingly diverges from it: 'And when I saw that he finally realised, in a moment of astonishment, that I was going to, I shot him.'[2] There is a touch of sadism here which is uncharacteristic of the mythic Hero. Grey also seems to have an Aristocratic passion for danger which clearly underlies his joy in flying and amateur jump riding. All Dick Francis's riders take danger for granted as a necessary concomitant of an activity they love for other reasons, and to some extent enjoy it, but for Grey it seems to be an essential ingredient of his pleasure.

The final movement of the book begins with Grey deciding to escape from his pursuers in a heavy DC-4 which he has no experience of flying, rather than the small Cessna which would have been no problem for him. Once airborne, he recognizes the Aristocratic impulse which has driven him: 'Nothing but pride. I was still trying to prove something to Billy, even though he was dead.... Childish, vainglorious, stupid, ridiculous: I was the lot.' 'Vainglorious' is the key word here; it associates his action with the 'ofermod', the excessive pride and self-confidence, which causes the leader of the Anglo-Saxon troop defending Maldon to give up his winning geographical advantage when the Vikings taunt him; *The Battle of Maldon* dramatises a moment when the Aristocratic principle disastrously takes over from the Heroic, not long before the Norman Conquest.

There is enough of the Hero in Henry Grey to make him bitterly self-critical of that Aristocratic folly, as the extract above shows, and Francis carefully structures the narrative so that Grey can subsequently discover a sound Heroic reason why the decision to take the DC-4 had been the right one after all, but the impulse had been Aristocratic, like most of Grey's impulses. What is very curious and engaging about all this is that Henry Grey is a social aristocrat; he actually inherits an earldom part-way through the story. But he is a deviant in Francis' work; with only this and one or two more recent protagonists as exceptions, his heroes, widely disparate as are their professions and social origins, are clearly mythic Heroes, as were the first great police detectives of Dickens and Collins, and as are some of the most popular of recent detectives, such as Inspector Wexford and Brother Cadfael.

These two myths balance each other well: the myth of the Aristocrat involves romanticism, glamour, pride and the glorification of death; the myth of the Hero celebrates reason, energy and the love of life. Exemplars of both set their own standards, but the mythic Aristocrat cannot forgive, and often cannot survive, falling below his, while the mythic Hero recognizes the necessity of accepting and starting again from his failures. One is the myth of paranoia and sadomasochism; the other the myth of maturity and sometimes rather self-righteous and even humourless sanity. Both have a certain element of extremity, and, above all, of certainty.

It was the novels of Henry James and George Meredith that first suggested that there was a third substantial myth of male character which seemed to become important to writers at a time when uncertainty was making particularly strong inroads into their mind-set. Characters like Ralph Touchett and Merthyr Powys seemed to exist on the margins of life, to be sensitive, attentive listeners, never quite fully engaged. Characters of this kind, and others whose relation to their society seemed to be neither the proud isolation of the Aristocrat nor the positive engagement of the Hero, but something more ambiguous, tentative and often absurd, seemed to account for many recurring types who could not be seen as either Aristocrats or Heroes. For them, it seemed to be necessary to accept certain constraints of society in order to elude others and retain a precious and precarious individuality, and this led to postures and gestures which struck either the characters themselves or those observing them – and sometimes both – as absurd. This observation led to the naming of this concept as the myth of the Clown.

At first sight, it looked to be an essentially modern myth, since there are few examples in the first half of the nineteenth century and the Clown seems to belong particularly well to the age of existentialism and the Theatre of the Absurd. But, as noted above, it seems now rather that this myth becomes of particular relevance in times when certainties are lost and men find it necessary to try on different rôles and to protect an individuality become newly precious to them, by making some contract with the society with which they are not wholly at ease – like the schoolboy 'class clown' who would otherwise be bullied. Analogues can be found during the sixteenth century, another age in which uncertainty was endemic. A great number of

Shakespearean characters, for example, besides those actually listed in the cast as clowns, are mythic Clowns: from obvious examples like Lear's Fool, and buffoons such as Bottom and Falstaff, to the subtler example of Hamlet, who, as Harry Levin long ago pointed out in *The Question of Hamlet*,[3] can be seen from time to time slipping into the unfilled shoes of his father's jester, Yorick.

If the mythic Aristocrat is set above society and the mythic Hero is thoroughly engaged with it, the mythic Clown is, in one sense or another, unabsorbed by it, sometimes caught on the outside with his face pressed sadly to the glass, sometimes, in the words of John D. MacDonald's mythic Clown, Travis MacGee, choosing 'to live free in the nooks and crannies of the huge and rigid structure of an increasingly codified society.'[4] Mythic Clowns are unlikely to be either solid businessmen or charismatic leaders, and they rarely feel they have a secure place in the world, though they often do not care. The myth of the Clown is a neurotic's myth, involving withdrawal, anxiety and self-mockery – 'God! your only jig-maker' – but also the myth of the mischievous and unregenerate child, the joker in the pack, the id to the Aristocrat's ego and the Hero's superego. As seems appropriate to an association with childhood rather than with adolescence or maturity, this myth is peculiarly fluid and transitional, like childhood, so that its taxonomy is more problematic than that of the myth of the Hero, with its single template, or the Aristocrat, where the types are quite sharply distinct.

Clowns shift easily from one type to another, and some even move to the aegis of another myth. There is a certain kind of thriller, dazzlingly practised by Michael Innes in novels like *From London Far* (1946) and *The Journeying Boy* (1949), that turns on the transformation of the central character from Clown to Hero under the pressure of circumstances, a kind of *Bildungsroman*, though the character being educated by events is likely to be middle-aged rather than young.

Like children, some mythic Clowns keep getting things wrong; like children, some mythic Clowns are always in trouble; like children, some mythic Clowns play tricks on the the pompous; like children, some mythic Clowns can spot when the emperor has no clothes; and like children, some mythic Clowns never stop asking questions. Such characteristics make clowns very appropriate detectives: in Tony Hillerman's *Sacred Clowns* (1993), a Native American character speaking of the clown-like *koshares* whose foolery is an essential part of a pueblo religious ritual, says:

> To outsiders they look like clowns and what they do looks like clowning. Like foolishness. But it is more than that. The *koshare*

> have another role. I guess you could say they are our ethical police. It's their job to remind us when we drift away from the way that was taught us... like policemen who use laughter instead of guns and scorn instead of jails.[5]

Every Clown template I have been able to recognize has at least one representative among series detectives. There are, as hinted initially, a number of these templates, and they cover a wide range of clownish styles. At one end of the scale, perhaps, is the Holy Fool, rarely found in detective fiction, but immensely important in certain kinds of serious literature. The Holy Fool is a little over the edge of sanity, but his oddity or insufficiency has a sacred quality; he has been 'touched' by the finger of God. The Fool in *King Lear* is the crucial example here: he seems, like a child, to say exactly what he thinks in spite of the constant risk of punishment and to get away with surprisingly outrageous utterances, the only free voice in a sick and split court. Other examples of the Holy Fool in literature are Don Quixote, who perceives the world quite differently from those around him, and Dostoevsky's Prince Myshkin, who is certainly not the idiot the usual translation of the novel's title suggests, but an epileptic with a childlike naivety which gives him a strangeness which other people find hard to tolerate. For writers, epilepsy seems a favourite metaphor for the strangeness of the Holy Fool; Simon in Golding's *Lord of the Flies* is another example.

There seems to be only one unequivocal detective Holy Fool and perhaps even he is not wholly unequivocal, since when, in the very first story about him (in a book appropriately named *The Innocence of Father Brown* from 1911), he throws soup at a wall, swops greengrocers' tickets about and generally plays the fool, it is with detective purpose. Nevertheless, there is a generally acknowledged quality of gentle absurdity about him which fits the template. It should perhaps be noted here that a religious is by no means necessarily a Holy Fool, or any kind of Clown: as noted above, Ellis Peters' Brother Cadfael is clearly a mythic Hero.

It had seemed that Inspector Ghote might qualify as a Holy Fool, but the analysis by Reginald Hill, in his essay on H.R.F. Keating for *Twentieth Century Crime and Mystery Writers*, neatly establishes a distinction:

> Doubts and fears and errors and awkwardnesses are so universal that nearly every culture has a place in it for the Holy Fool whose very simplicity can open up the way to truth and understanding, or perhaps more to the point (for Ghote is no fool and not very holy)

> some fable or parable in which a quiet, self-effacing and disregardable man is forced out of his diffidence by duty.[6]

No fool, not very holy, quiet, self-effacing, diffident: this is the template which, taking the great Charlie Chaplin example as paradigmatic, must be called the Little Man, both funny and pathetic, an apparent loser who sometimes triumphs. This Clown is the one most noticeably outside society with his face pressed to the window, not quite able to understand his exclusion. Ghote is an admirable example, as can be seen in *Inspector Ghote Goes by Train* (1971), where Ghote's carriage gradually fills with a remarkable collection of mythic Clowns of various types, most notably the Demonic Buffoon – of which type much more will be said – whom he is trying to catch. The Little Man is surprisingly common as a detective, particularly in the second half of the century. The television police detective Colombo, with his shabby raincoat and apologetic manner, never so apologetic as when he is about to entrap a murderer, shows how useful can be the contempt a Little Man attracts from the rich and vicious, to an investigator trying to establish their guilt.

With Ghote, Colombo and Arthur Upfield's Aboriginal Boney in mind, it might be thought that the Little Man detective is essentially a member of what in this country counts as an ethnic minority, with outsider status conferred by membership of that minority; Reginald Hill's Afro-Caribbean Joe Sixsmith is a recent example. But this is not necessarily so. A powerfully discomfiting example of a Little Man is Peter Dickinson's Pibble, whose very name is clownish. An ageing Detective Superintendent, 'unglamorous, greying towards retirement', Pibble is first met in *Skin Deep* (1968), where he makes a disastrous mistake about the identity of the killer. Here the most poignantly clownish moment comes when he holds out a pair of shattered National Health spectacles 'consolingly' to one who had loved the murderer, beginning to say, 'Crippen [his favourite expletive], I'm – ' – he would have said 'sorry' but is struck down in the middle of the inane inadequacy. This is clownish not only in the sense of being foolishly inept, but, still more relevantly, because it expresses his incomprehension of the passionate grief of which the person he is addressing is capable. His own emotional life is a desert; he and his wife have lived together for many years but are completely unacquainted. It seems appropriate that in the remarkable final novel of the series, *One Foot in the Grave* (1980), he begins in a state of premature senility, completely touched.

Pibble's mistake in *Skin Deep* almost puts him into another Clown class: the Gull. This naive person, ripe for plucking, is one of the great traditional clownish rôles: see, for example, the shepherd's son in *A Winter's Tale*. In the detective story, his main function is to get things wrong along with the reader. First person narrative can be a problem for the detective story writer, since if readers are privy to all the detective's thoughts, they may solve the mystery too soon. The first way of solving this problem was to use the detective's associate as narrator – Gabriel Betteridge, or Watson and the other idiot associates who stumble along in his wake, or Archie Goodwin, trusted but kept partially in the dark so that his employer can astonish him; a more modern technique is to use the narrator/detective as Gull. In the novels beginning, aptly enough for my theme, with *Puzzle for Fools* (1936), by the pair writing as Patrick Quentin, it is almost axiomatic that the person confided in most freely by the theatrical producer and Gull, Peter Duluth, will turn out to be the murderer. Not a first person narrator, Simon Brett's Charles Paris is another Gull, although he tends to be taken in more by his agent and the sharper theatrical types who surround him than by the murderer.

The naivety of the Gull is genuine and goes all through; that of many other Clown detectives is partially a disguise, like the innocence of Father Brown, or the manner of another type, the detective Silly Ass; this rôle was probably invented by Baroness Orczy, when she made the absurdly dandified Sir Percy Blakeney the alter ego of the heroic Scarlet Pimpernel, though there are, of course, mythic Silly Asses in earlier periods, especially in Restoration Comedy's Flutters and Witwouds, and P.G. Wodehouse perfected the style of the type. Both Dorothy L. Sayers and Margery Allingham initially dressed their detectives as Silly Asses, although, demonstrating the fluidity of the Clown, Lord Peter Wimsey, as noted above, develops into a Perfect Aristocrat.

As for Campion, Margery Allingham has remarked that he established himself, against her original intentions, as the central character in her books, initially 'a rather foolish looking young man' used for comic effect.[7] In an early novel, *Look to the Lady* (1931), he concludes a bizarrely delivered invitation to the heir of an aristocratic family facing a dire emergency with the offer of 'little pink cakes'. However, as her novels moved from adventure to pure mystery, he grows into another, more sober Clown type, the Lay Analyst. In the reversion to adventure of *Traitor's Purse* (1941) he behaves blindly as a Hero, but this is only a temporary metamorphosis. Recently, the mantle of a detective Bertie Wooster has been assumed with some comic effect by an American, and Lawrence Sanders' Archy McNally has all the classic mannerisms of the Silly Ass transported to Florida.[8]

The Lay Analyst is the least obviously clownish of all the mythic Clowns, a lonely, self-critical, self-conscious being, one who 'thinks too precisely on the event', as Hamlet does, though he is usually a better listener than Hamlet, receptive, responsive, concerned, the confidant of those more intensely involved in life than he is himself. Sometimes, as H.R.F. Keating observes of the later Campion, perhaps a little harshly, he is 'little more than a pair of observing eyes'.[9] A comparison of Ross Macdonald's Lew Archer with Raymond Chandler's Philip Marlowe clarifies the type. Marlowe, while a degree less tough, ruthless and self-sufficient than Dashiell Hammett's Sam Spade, is nevertheless a mythic Hero, more doer than dreamer, in spite of the metaphoric richness of the narrative prose Chandler endows him with. Lew Archer, though capable of tough action at need, is essentially a private ear, someone in whom the characters of the novel confide as if he were their psychiatrist; it is his sympathy, rather than his toughness, which solves the puzzles.

Robert B. Parker has taken a further step with the literate ex-boxer, Spenser, a Lay Analyst in a long-term love relationship with a professional analyst, as well as a mythic friendship with Hawk, a black Satanic Aristocrat. He and Albert Campion are two examples of the type which contradict an observation valid for many mythic Clowns, that they are likely to be solitaries with lost, unhappy or unachieved relationships. As well as his profound relationship with Lady Amanda Fitton, which in *Traitor's Purse* is almost lost through clownish diffidence, we see Campion constantly accompanied by a Buffoon, his manservant Lugg, to execute the cruder work. There is no doubt of his own clownish credentials, however, as his resistance to being sent out to govern New South Wales, as it were, demonstrates in *More Work for the Undertaker* (1948). This marks his self-chosen rejection of the constraints, disguised as opportunities, of the non-mythic aristocratic society into which he was born.

Another archetypal Lay Analyst with a good marriage is Ngaio Marsh's Roderick Alleyn, who may seem at first sight to be some kind of mythic Aristocrat, with his often mentioned 'profile of a Spanish grandee'. Perhaps the fact that this is a foreign, not an English aristocratic look, unlike that of yellow-haired Lord Peter Wimsey, prevents misapprehension. As an English gentleman, and a particularly particular one, fastidiously troubled by his sergeant's taste for pickles, he is clearly on the outside of the police culture in which he works, while his choice of profession has similarly put him outside the culture to which he was born. But it is also this choice of profession, one in which a pact must be made with society, that marks him as no mythic Aristocrat, but a serious mythic Clown.

A later, and to my mind less pleasant, though perhaps more complete example is P.D. James's poet-policeman, Adam Dalgliesh.[10] That he is on the outside of comfortable humanity is stressed by the references to the loss of his wife and child, and his inability to begin a new relationship with any of the women, such as Cordelia Gray, put in his way by the author; Superintendent Bone, the Lay Analyst detective of Staynes and Storey, is also said to have lost his first wife in distressing circumstances,[11] but their novels have traced a new relationship as far as incipient fatherhood for him, and he has nothing of Dalgliesh's Pierrot Lunaire coldness.

To move from a Dalgliesh to a Dalziel is to leap from Lay Analyst, the reflective Pierrot, to Buffoon, the truly comic mythic Clown. Buffoons subdivide into two templates: the more powerful Bouncer, like Falstaff, who is outside society mainly because he will not hold still for long enough for it to incorporate him, and the less powerful Prankster, like Autolycus, who shares some characteristics of the Little Man. A great many comic duos have made their stock-in-trade the interaction of the two. The *locus classicus* must be Laurel and Hardy (though they have an honourable ancestry in Sir Toby Belch and Sir Andrew Aguecheek) but there are many other pairs in which one partner is large, rotund and forceful, while the other is small and apparently vulnerable, but somehow, often through a real or affected stupidity, capable of embroiling the other in difficulties which, though he shares and often suffers more severely from them, nevertheless give him a certain *Schadenfreude*. Buffoons do not always come in pairs like this, but they are always more or less funny, where Holy Fools and Lay Analysts are often wholly serious, more like the sad Pierrot of *Les Enfants du Paradis*.

Buffoons may not at first sight seem well adapted to be detectives. Irresistibly one is reminded of Inspector Clouseau, an indomitable and hopeless Buffoon. It must be remembered, however, that the structure of comedy of the Pink Panther kind insists that the detective triumphs, however accidentally. Sometimes by inadvertence, sometimes by being much sharper than he is given credit for (as a child may be), the Bouncer bounces home, the Prankster emerges triumphant from a cloud of confusion.

The distinction between the two versions of Buffoon is neatly made by comparing Reginald Hill's Dalziel with R.D. Wingfield's Frost. Both have been recently well embodied on television, though Warren Clarke's superbly felt Dalziel does not have quite enough of a beer gut.

Dalziel is quite successful in spite of the frequency with which he oversteps one mark or another, progressing to Detective Superintendent as Hill's sequence proceeds. He is capable, in *Child's Play* (1988), of manipulating events to the point where a despised and unsuitable superior is kept out of a coveted job, and then, in *Under World* (1989), of rescuing the man from difficulties of his own making. In *Bones and Silence* (1990) he actually gets to play God in a medieval mystery play. This is the bigness of the Bouncer.

Frost, on the other hand, is set up to be kept an inspector all his life by the malice of his superior and the kind of revenge he practises is, for example, to put the man into the position of giving a lift in his elegant new car to an old lady Frost knows to be incontinent. Add to this his trademark scarf and the raincoat he seems to have borrowed from Colombo and his close kinship with the Little Man becomes evident; his is the scale of the Prankster.

By no means all Buffoon detectives are police, however; there are splendid examples of both Bouncer and Prankster in the private sector. The American Anglophile who wrote as John Dickson Carr and Carter Dickson created two great Bouncers in Dr Fell and Sir Henry Merivale. This writer's tales, with the disturbing exception of *The Burning Court* (1937) are dedicated to restoring sense and reason to a world which appears to have become insane – the true *raison d'être* of the locked-room mystery. Dr Fell's massive energy and cheerfulness are vital to the sense-making in the novels in which he appears, but clown-like, he is not altogether a part of the ordinary world whose sense he restores. This point is made strongly in an early story, *Hag's Nook* (1933), when Fell is telling Ted Rampole, the young American through whose eyes the story is seen, about the dark history of the Starberth family: 'In the restaurant-car, this swilling, chuckling fat man had seemed as hearty as an animated side of beef; now he seemed subdued and a trifle sinister.'[12] For a moment, Dr Fell himself seems like one of the hobgoblins he is so often engaged against and we see that he might be a strange creature indeed; like the eponymous character in A.A. Milne's play *Mr. Pym Passes By*, first seen some 14 years earlier, this genial and yet alarming character, full of old English lore, may be a version of Robin Goodfellow. And this seems right: the Old People, whether Shakespeare's or Kipling's, certainly belong to the myth of the Clown; when Bottom, an unquestionable Bouncer, falls into the company of Titania and her attendants, he is visiting his distant cousins.

Sir Henry Merivale, in the stories signed Carter Dickson, is a less ambiguous Bouncer, the great lord of misrule. He shows, perhaps more clearly than any other example, the eternal child in the myth of the Clown. Carr/Dickson, an American romantically in love with England, based H.M. very loosely on Sir Winston Churchill, as he had based Fell more closely on Chesterton, and it was perhaps Churchill's physical appearance, especially in his siren suit, as of a giant baby, which suggested the monumental and often engaging childishness which gets H.M. into so many ludicrous situations. Examples at random: in *The Skeleton in the Clock* (1949), he disrupts an auction by sticking a pompous grande dame in the backside with a pike; in *Night at the Mocking Widow* (1951), he dresses up as an Indian chief for a village fete and brings the affair to a conclusion with all the village children, and a good many of their elders, hurling mud pies; in *She Died a Lady* (1943) he drives a motorised wheelchair at breakneck pace through one of the sleepy English villages Carr/Dickson so loved, to the total destruction of decorum and the partial destruction of the village inn. As R.E. Briney points out in *Twentieth-Century Crime and Mystery Writers*, these outbreaks of mischief are rarely pointless or merely farcical elements in the novel's construction.[13] In the last-mentioned, for example, the initial mad wheelchair ride prepares the way for a second and that, leaving H.M. suspended half over the edge of a cliff, is deeply suggestive of the way the lovers had managed their disappearing trick. When *Seeing is Believing* (1941) is punctuated by H.M.'s dictation of memoirs consisting almost entirely of the record of practical jokes played on a hated uncle, the reader is warned to take another look at an innocent-seeming uncle in the main plot.

Like Dalziel, these other Bouncers are men of substance – in more than one sense – and have a certain amount of power. H.M. indeed is chief of a counter-espionage department, though about to be kicked upstairs to the Lords. Pranksters, on the other hand, tend to be permanently impoverished and to have only the limited power their wits give them in particular situations. Jonathan Gash's Lovejoy, for example, is always poor, always dominated by the rich, gorgeous and assertive women he seduces and exploits, always claiming terror and yet always triumphant over the murderous villains he encounters, often by playing some dire trick on them. A prime example is his escape from and contribution to the Edinburgh fairground rumble in *The Tartan Ringers* (1986). Mike Ripley's Angel – although, like Campion and Alleyn, he has aristocratic connections – is another example.[14] More recently there has been Marcus Didius Falco, Lindsey Davis' Ancient Roman private eye.[15] He is bullied by his mother, his sisters and his inamorata Helena, always fending off his landlord, unable to win the wealth he needs to become a Senator and thus entitled to marry Helena – and yet, like Lovejoy, he always bounces back from whatever disaster befalls him and always outwits the enemy by devious means.

It will be noticed that all these prankster detectives are recent creations; it is hard to find an example earlier than Malone, the 'little Irish lawyer' in Craig Rice's detective stories of the late 1940s and early 1950s,[16] though it is possible that he owes something to Old Sharon in Wilkie Collins' *My Lady's Money* (1878). Malone, however, was soon to be followed by Abdel Arthur Simpson, anti-hero of two novels by Eric Ambler, *The Light of Day* (1962) and *Dirty Story* (1967): a minor con-man, illegitimate, desperate to get a British passport, terrified of the dangerous people he is associating with, but a survivor. These were rare birds, but now fellows cut from this template are everywhere. The series detective certainly began in the myths of the Aristocrat and the Hero and the attraction of the grand simplicities of the latter is perhaps demonstrated by the regular appearance at the head of the best-seller lists of Dick Francis and, until her death, of Ellis Peters. But the Clowns have been gaining ground ever since the 1930s, the Age of Anxiety. That so many series detectives are now mythic Clowns of one kind or another may have something to say about an increased sophistication in the readership, able to respond to the more demanding fluidity and ambivalence of the myth of the Clown. Perhaps the particular efflorescence of the Prankster in recent years suggests something further: that the ducking and weaving characteristic of this slightly fragile Buffoon represents something that many readers unconsciously recognize as necessary to preserve one's selfhood in the teeth of the constraints and pressures of the fag-end of the twentieth century.

Notes

1 Rex Stout, *The League of Frightened Men* (London: Cassell, 1935) p. 16.
2 Dick Francis, *Flying Finish* (1966; repr. London: Pan, 1968), p. 195.
3 Harry Levin, *The Question of Hamlet* (New York: Oxford University Press, 1959), pp. 121–6.
4 John D. McDonald, *The Quick Red Fox* (1964; London Edition: Robert Hale, 1966), p. 113.
5 Tony Hillerman, *Sacred Clowns* (London: Michael Joseph, 1993), pp. 164–5.
6 Lesley Henderson (ed.), *Twentieth Century Crime and Mystery Writers*, 3rd edn. (Chicago and London: St James Press, 1991), p. 617.
7 See Richard Martin, *Ink in her Blood: The Life and Crime Fiction of Margery Allingham* (Ann Arbor and London: UMI Research Press, 1988), p. 57.
8 See Lawrence Sanders, *McNally's Secret* (1992), *et seq.*.
9 Henderson (ed.), *Twentieth Century Crime and Mystery Writers*, p. 19.

10 See P.D. James, *Cover her Face* (1987), *et seq.*.
11 See Jill Staynes and Margaret Storey, *Goodbye Nanny Gray* (1987), *et seq.*.
12 John Dickson Carr, *Hag's Nook* (New York and London: Harper & Brothers, 1933), pp. 20–1.
13 Henderson (ed.), *Twentieth-Century Crime and Mystery Writers*, pp. 177–8.
14 See Mike Ripley, *Angel City* (1994), and so on.
15 Lindsey Davis, *The Silver Pigs* (1990), *et seq.*.
16 See Craig Rice, *Trial by Fury* (1950) and so on.

9
Mean Streets and English Gardens

Warren Chernaik

I

It has often been argued that detective fiction is an intrinsically conservative form. At the narrative's inevitable, carefully foreshadowed conclusion, the mysteries are solved, the elements of fear, suspicion and doubt are banished, order is restored: readers come to detective fiction expecting to find reassurance, to have their formal and ideological expectations confirmed. As Franco Moretti has said:

> One reads only with the purpose of remaining as one already is: innocent. Detective fiction owes its success to the fact that it teaches nothing.[1]

According to Moretti, Dennis Porter and other influential theorists of the genre, who like to quote Foucault's paradigm of the all-seeing Panopticon, providing universal surveillance to keep potentially disruptive forces under firm control, 'detective fiction is a hymn to culture's coercive abilities.' In Porter's view, 'works in the genre always stand in defense of the established social order', even where they 'uncover corruption among prominent citizens and public officials'.

> The point of view adopted is always that of the detective, which is to say, of the police, however much of an amateur the investigator may appear to be. In a detective story the moral legitimacy of the detective's role is never in doubt.[2]

And yet in much of the most interesting detective fiction, English and American, it is precisely the 'moral legitimacy of the detective's role', as well as that of the forces of order in the state and in society, which is called into question. The novels of Agatha Christie are often characterized as conservative and class-ridden, comforting in their predictability, presenting an idealized landscape shaped to fit the prejudices of their middle-class audience: 'Agatha Christie's world', one critic writes, 'was never more than nostalgia and illusion'.[3] Raymond Chandler, dismissing the 'Golden Age' tradition of the formal detective story, finds such works unsatisfactory both in 'the artificial pattern required by the plot' and in their ideology. Dedicating a volume of his short stories to the editor of *Black Mask*, Chandler writes of 'the time when we were trying to get murder away from the upper classes, the weekend house party and the vicar's rose garden, and back to the people who are really good at it', and in 'The Simple Art of Murder' he praises Dashiell Hammett as a realist impatient with received ideas:

> Hammett gave murder back to the kind of people who commit it for reasons, not just to provide a corpse; and with the means at hand, not with hand-wrought duelling pistols, curare, and tropical fish.[4]

And yet, as Alison Light has shown, the detective novels of Christie owe much of their appeal to the way they challenge rather than reinforce the assumptions of readers, both about the genre and about the society depicted. In a Christie novel, no one is to be trusted (even the Watson-like narrator, in *The Murder of Roger Ackroyd*); aside from the detective, each character is both a potential murderer and potential victim. The subject of a typical Christie novel (and the same generalization can apply to many of the novels of P.D. James, an author consciously working in the same tradition) is 'the unexpected violence which manifests itself ... in that most apparently secure of places, family life', as in the course of the narrative, 'a safe, known world', familiar in its contours, is suddenly 'thrown out of kilter':

> For Christie, as her characters must always realise with alarm, the criminal is first of all *one of us*, someone who for nine-tenths of the novel must carry on seeming successfully to be just that. ... The most innocent (the least likely) person may turn out to be the criminal. ... It is within the charmed circle of *insiders* that the criminal must be sought.[5]

W.H. Auden has argued in 'The Guilty Vicarage' that much of the appeal of the detective story, which he likens to 'an addiction like tobacco or alcohol' ('once I begin one, I cannot work or sleep till I have finished it') is based on 'the dialectic of innocence and guilt'.

> I suspect that the typical reader of detective stories is, like myself, a person who suffers from a sense of sin.... The magic formula is an innocence which is discovered to contain guilt; then a suspicion of being the guilty one; and finally a real innocence from which the guilty other has been expelled.... The fantasy, then, which the detective story addict indulges is the fantasy of being restored to the Garden of Eden, to a state of innocence, where he may know love as love and not as the law. The driving force behind this daydream is the feeling of guilt, the cause of which is unknown to the dreamer.[6]

C. Day Lewis, the author of several fine detective novels under the pseudonym Nicholas Blake, similarly characterizes detective fiction as an 'outlet for the sense of guilt' in its readers.

> The pattern of the detective-novel [is] as highly formalized as that of a religious ritual, with its initial necessary sin (the murder), its victim, its high priest (the criminal) who must in turn be destroyed by a yet higher power (the detective).... The devotee identified himself with both the criminal and the detective representing both the light and dark sides of his nature.[7]

John Cawelti, following Auden, defines the psychological process underlying the classical detective story as 'the fantasy projection of guilt away from the reader', arguing a variant of the thesis that detective fiction is essentially conservative in its ideology:

> instead of laying bare the guilt of bourgeois society the detective–intellectual uses his demonic powers to project the general guilt onto specific and overt acts of particular individuals, thus restoring the serenity of the middle-class social order.[8]

In aesthetic terms (leaving aside questions of ideology), the recurrent pattern described by Auden and others is akin to the Aristotelian catharsis: emotions are aroused in the reader in the course of a work in order to be purged at the end, restoring a sense of balance, physical and emotional health, as the work reaches its preordained conclusion.

Yet there is always a residue of unease at the end of most novels of crime and detection, whether the author is Christie, Dorothy L. Sayers, P.D. James, Dashiell Hammett, or Walter Mosley. Guilt can never be entirely dissipated, the criminal Other never cordoned off, the potential for the sudden eruption of that which is most feared brought wholly under control. From its origins in Poe and Conan Doyle, detective fiction has always been 'a genre in which appalling facts are made to fit into a rational or realistic pattern', in which, as Ross Macdonald says in his splendid essay 'The Writer as Detective Hero', 'the nightmare can't quite be explained away, and persists in the teeth of reason':

> An unstable balance between reason and more primitive human qualities is characteristic of the detective story. For both writer and reader it is an imaginative area where such conflicts can be worked out safely, under artistic controls.[9]

Detective fiction cannot, then, be accurately categorised as ideologically conservative: there is too much variety within a genre which has attracted such authors as Poe, Ruth Rendell, Sara Paretsky, Reginald Hill, Michael Dibdin, and Josef Škvorecký. Detectives are shown as fallible, fatally compromised, unable to disentangle themselves from a corrupt society: detectives fail to solve a crime because they are unable to rid themselves of certain prejudices or ideological assumptions (*Trent's Last Case*, Reginald Hill's *Deadheads*); they solve the crime, but in an access of sympathy or revulsion allow the criminal to go free (*An Unsuitable Job for a Woman*); they wade knee-deep in blood, like the hero of a Jacobean revenge tragedy (Hammett's *Red Harvest*). In several of Michael Dibdin's detective novels, the police and the state apparatus are shown to be so corrupt that the detection and punishment of crime become virtually impossible; in James Ellroy's novels, criminals and police are indistinguishable. In the Lieutenant Boruvka stories of Škvorecký, the flawed, unprepossessing detective, struggling to maintain his integrity, is impeded at every turn by a dishonest, self-seeking bureaucracy: the police here are servants of an unjust authoritarian state unremittingly hostile to the truth the detective seeks, and in a bitter coda, *The End of Lt Boruvka,* the hero, finally rebelling against the contradictions of his rôle, is himself sent to prison.

But one point on which practitioners of the classic Golden Age novel and the hard-boiled school agree is that the detective novel is formally conservative. Detective fiction presents 'a world without chance' in which everything is planned, yet artfully disguised by an author at all times in control.

> The skill of the detective author consists in being able to produce his clues and flourish them defiantly in our faces: 'There!' he says, 'what do you make of that?' and we make nothing.[10]

One indication of the role in detective fiction of the 'hermeneutic game', the puzzle set as challenge to the reader, is its tendency to codify, setting forth rules of the game (generally stated in terms of what to avoid): Ronald Knox's 'A Detective Story Decalogue', S.S. Van Dine's 'Twenty Rules for Writing Detective Stories', John Dickson Carr's 'Locked-Room Lecture'. The initiation ceremony of the Detection Club of London, founded by G.K. Chesterton and Anthony Berkeley in 1928, includes the oath 'Do you solemnly swear never to conceal a vital clue from the reader?'[11] The insistence on form as the one over-riding imperative – the view that 'the distinguishing characteristic of a detective story, in which it differs from all other types of fiction, is that the satisfaction it offers to the reader is primarily an intellectual satisfaction', or that 'there is no essential difference between a detective novel and a mathematical puzzle' – has led some theorists to define the genre in Procrustean terms. Auden, for example, denies that the novels of Raymond Chandler can be classified as detective fiction.

> Actually, whatever he may say, I think Mr Chandler is interested in writing, not detective stories, but serious studies of a criminal milieu, the Great Wrong Place, and his powerful but extremely depressing books should be read and judged, not as escape literature, but as works of art.[12]

Yet the primacy of plot, the emphasis on authorial control and the shaping of a narrative, the careful parcelling out of information to the reader, mystification with the promise of eventual enlightenment, are characteristic of detective fiction of all schools. Dorothy L. Sayers, contesting the view that the categories of 'work of art' and detective novel are mutually exclusive, sees the formal patterning, the necessary closure, of detective fiction as one of its greatest attractions:

> There is one respect, at least, in which the detective-story has an advantage over every other kind of novel. It possesses an Aristotelian perfection of beginning, middle, and end. A definite and single problem is set, worked out, and solved; its conclusion is not arbitrarily conditioned by marriage or death.[13]

Many writers of detective fiction, differing widely in other respects in their approach to the genre, have expressed similar views. Minette Walters in an interview remarks 'what is so good about it is that it has a very tight structure. People who write general fiction often flounder because the structure is far too loose, and they lack the tightness that genre fiction imposes on you', while Raymond Chandler in a letter deploring the tendency to dismiss such novels as 'easy reading' argues: 'their form imposes a certain clarity of outline which is only found in the most accomplished "straight" novels.'[14] Nearly all detective stories follow Poe's advice and are written backwards, with details carefully calibrated, clues planted, to lead up to the conclusion from which the author begins. In Alison Light's words, applicable not only to the typical Christie novel but to the genre as a whole, the aesthetic principle on which detective fiction is based is 'a temporary fragmentation' in which 'the pieces are deliberately scattered for the express purpose of putting them back together again.'[15] Ross Macdonald is another writer who consistently held that 'the limitations of popular art can be liberating' and that working within such limitations suited him as an artist:

> I think that the overall intention of a book is more important than the individual scenes, and the individual scenes have to reflect that intention. In other words, I believe in the principle of narrative unity.
>
> I see plot as a vehicle of meaning. It should be as complex as contemporary life, but balanced enough to say true things about it. The surprise with which the detective novel concludes should set up tragic vibrations which run backward through the entire structure. Which means that the structure must be single, and *intended.*[16]

A number of theorists of detective fiction, like Auden, have sought to draw a sharp distinction between 'escape literature' and 'works of art' or serious literature. Tzvetan Todorov, for example, in 'The Typology of Detective Fiction', writes:

> Detective fiction has its norms; to 'develop' them is also to disappoint them: to 'improve upon' detective fiction is to write 'literature', not detective fiction. The whodunit par excellence is not the one which transgresses the rules of the genre, but the one which conforms to them.[17]

Where structuralists like Todorov and John Cawelti are professedly non-evaluative in distinguishing 'literature' from genre fiction, the more traditional critics of an earlier generation represented in *The Art of the Mystery Story* tend to use terms like 'literature' or 'art' to praise one category of text and denigrate another. Edmund Wilson in his amusing 'Who Cares Who Killed Roger Ackroyd?' sternly dismisses detective fiction as falling short of the highest literary standards, unremittingly second-rate, 'degrading to the intelligence', while other critics, like Marjorie Nicolson and Joseph Wood Krutch, use detective fiction as a stick to beat experimental modernism, praising detective fiction for its unpretentiousness, its 'welcome objectivity' and straightforward narration, seeing it as 'escape not from life, but from literature'.[18] Dorothy L. Sayers is strikingly inconsistent in her assessment of the genre she chose to work in. On the one hand, she anxiously defends its 'artistic status' – 'the detective story may be a highly finished work of art, within its highly artificial limits' – while on the other hand, in the same essay, she relegates it to the foothills of Parnassus because of its limited emotional range:

> It does not, and by hypothesis never can, attain the loftiest level of literary achievement. Though it deals with the most desperate effects of rage, jealousy, and revenge, it rarely touches the heights and depths of human passion.... A too violent emotion flung into the glittering mechanism of the detective-story jars the movement by disturbing its delicate balance.
>
> The detective story is part of the literature of escape, and not of expression. We read tales of domestic unhappiness because that is the kind of thing that happens to us; but ... we fly to mystery and adventure because they do not, as a rule, happen to us.[19]

Chandler, in his celebrated essay 'The Simple Art of Murder', explicitly rejects the distinctions Sayers is arguing here, and both in that essay and in several letters is scathing about the kind of belle-lettristic criticism of detective fiction represented in *The Art of the Mystery Story*. 'Neither in this country nor in England', he writes in a letter, 'has there been any critical recognition that far more art goes into these books at their best than into any number of fat volumes.' In another letter, commenting on the 'so-called critical essays on the detective novel' in the Haycraft collection, he complains, 'there is a constant haste to deprecate the mystery story as literature for fear the writer of the piece should be assumed to think it important writing.'

> I do not know what the loftiest level of literary achievement is: neither did Aeschylus or Shakespeare; neither does Miss Sayers. Other things being equal, which they never are, a more powerful theme will

> provoke a more powerful performance.... As for literature of expression and literature of escape, this is critics' jargon, a use of abstract words as if they had absolute meanings. Everything written with vitality expresses that vitality: there are no dull subjects, only dull minds.[20]

But in certain important respects Chandler and Sayers were in agreement, however much they may have diverged in their approach to the appropriate style and subject matter of detective fiction. Both were dissatisfied with the view which equated the genre with a 'crossword puzzle', caricatured by Sayers as a belief 'that every vestige of humanity should be ruthlessly expunged from the detective novel.'[21] Though Sayers preferred to work within generic traditions which Chandler sought to redefine radically, both writers had wider ambitions for the detective novel. Sayers used the Arnoldian phrase 'criticism of life', suggesting that the mechanics of plot construction are less significant, for readers as well as writers, than a recognition that a particular text treats 'a subject about which [the author] really had something to say'.

> If the detective story was to live and develop it *must* get back to where it began in the hands of Collins and Le Fanu, and become once more a novel of manners instead of a pure crossword puzzle.[22]

Though Chandler disliked the phrase 'novel of manners', he shared Sayers's view that the emphasis in detective fiction must be 'on people, not on facts', that the distinction between 'serious novel' and popular fiction was otiose and that the ideal to aim for should be 'a novel... which, ostensibly a mystery and keeping the spice of mystery, will actually be a novel of character and atmosphere.'[23]

Both Chandler and Ross Macdonald gave Dashiell Hammett the credit for opening up new possibilities for detective fiction. To Chandler, paying tribute in 'The Simple Art of Murder', Hammett 'demonstrated that the detective story can be important writing'. Hammett's innovations, his successors Chandler and Macdonald consistently emphasized, were to a considerable extent a matter of style, the discovery of an appropriate language for a narrative which 'had a basis in fact' and was 'made up of real things'.[24] Critical commentary on Hammett and Chandler gives full recognition to the achievements of the two writers in inventing a demotic American style, based on 'the speech of common men' (or 'the language of the street') but consciously shaped and honed down, relying on laconic understatement, as in the early fiction of Hemingway. Chandler was explicit about his artistic aims – 'insofar as I am able I want to develop an objective method', he said in an early letter – and associates himself with Hammett in seeking transparency: 'He has style, but his audience didn't know it, because it was in a language not supposed to be capable of such refinements.'[25]

And yet Chandler never considered stylistic innovation an end in itself. In a letter commenting on his early stories written for *Black Mask*, he described what he 'cared about' most in his writing, and what he expected readers to remember, as 'the creation of emotion through language and description'.[26] Chandler saw his own writings and those of Hammett as experimental attempts to expand the boundaries of detective fiction and to 'make an art' out of materials and language a popular audience 'can understand'. In one letter, Sayers receives qualified praise as sharing some of these aims.

> I am engaged and have always from the beginning been engaged in the effort to do something with the mystery story which has never been done. In one way Hammett came close; in another way Sayers came close. Neither was capable of imparting emotions to the right nerves.

In 'The Simple Art of Murder' he praises Hammett as a kind of liberator, through whose discoveries later writers might be enabled 'to say things he did not know how to say or feel the need of saying'. Again the emphasis is on possible new directions, not always predictable, and on the falsity of distinctions between 'popular' and 'serious'.

> He did over and over again what only the best writers can ever do at all. He wrote scenes which seemed never to have been written before.... Once a detective story can be as good as this, only the pedants will deny it *could* be even better.[27]

As Macdonald points out in his critique of his two chief predecessors, the novels of Hammett and Chandler, though sharing certain conventions and techniques (and a common origin in the pulp fiction of *Black Mask*), differ in several fundamental respects. Macdonald contrasts the 'ferocious intensity' of Hammett, in novels like *Red Harvest* and *The Maltese Falcon*, where the central character, as well as the other characters, is treated 'from the outside, without affection, perhaps with some bleak compassion', with the more romantic vision of Chandler, in which 'the hard-boiled mask' conceals the author's 'poetic and satirical mind'. If the achievement of Hammett and of Chandler after him was 'to raise the crime story into literature', the technical means which made this possible was the discovery of a style, objective and apparently impersonal, which was capable of communicating an individual vision of life. In Macdonald's words

> Hammett was the first American writer to use the detective-story for the purposes of a major novelist, to present a vision, blazing if disenchanted, of our lives.[28]

Chandler and Macdonald both were conscious of working within a particular genre through which they could express their deepest concerns, portray a particular society in vivid, evocative language, create characters and milieu. The constraints of the genre, both writers felt, could liberate the imagination. In Chandler's words:

> What I have tried to do (and if I fail, someone will succeed) is not at all to get to be a good enough writer to do a 'straight novel'; I could have done that long ago, just as I could have trained myself to write slick serials, if I had really worked at it. The thing is to squeeze the last drop out of the medium you have learned to use.[29]

Much the same point is made by a later writer in the hard-boiled tradition, Walter Mosley, in a 1993 interview, paying tribute to Hammett and Chandler while indicating the ways in which he reinterpreted the genre to suit his own particular concerns. In Mosley's view, 'the genre itself is in flux, it's changing': his novels, like the detective novels of Chandler and Macdonald, are first-person narratives set in southern California, but depict a very different world. Mosley tells the interviewer that he met little success with his initial attempts at 'straight' fiction ('the publishers weren't interested') and then found he could reach an audience and say what he wanted to say in a detective novel: 'So what I was able to do in the mystery was to pull people in who are interested in the genre and still talk about the lives of Black people.' To Mosley as to Chandler and Macdonald, an artist can be wholly individual while working within the conventions appropriate to a particular genre. In this respect, he tells the interviewer, detective fiction and 'straight fiction or literature' were alike.

> No, I don't see any difference in it. Of course, in the genre there are certain kinds of things that you have to do, but it's the same in a

> coming-of-age novel, somebody has to come of age. So you have to follow the conventions. Good fiction is in the sentence and in the character and in the heart of the writer. If the writer is committed to and in love with what he or she is doing, then that's good fiction.[30]

Macdonald's intricately plotted, complex novels structured around a detective narrator engaged in 'putting together the stories of other people's lives and discovering their significance', are, like those of Mosley a generation later, explicitly attempts 'to bring this kind of novel closer to the purpose and range of the mainstream novel'. As Macdonald says:

> The basic idea of a detective story is to create a puzzle and provide a solution. But I also want to give the reader a full novel at the same time. I am not content unless I am giving myself and the reader everything in my capabilities, this is something an artist wants – to give everything he has to the work at hand. A detective novelist, like any other novelist, is trying to give a true sense of human life as he sees it.[31]

II

A strong element of social criticism links the novels of Hammett, Chandler, Macdonald, Mosley and Sara Paretsky, all working in the hard-boiled tradition. In the later writers, this aspect of the novels tends to be more overt. Paretsky's detective novels are consciously written from a feminist perspective and comprise a running critique of many of the assumptions of the hard-boiled genre, with its predatory *femmes fatales* and its cult of toughness and endurance.[32] The choice of a female private eye – or in Mosley's Easy Rawlins, a black detective – as the centre of consciousness in a novel in itself entails a perspective on the inequalities and contradictions of American society which differs substantially from that found in *The Long Goodbye* or *The Big Sleep*. Mosley shares Paretsky's revisionist approach to the genre, expressing in an interview his dissatisfaction with a tradition which he sees potentially as a vehicle for anatomizing 'the world we live in':

> The original characters in 'hard-boiled detective' fiction were a kind of spirit, the sad spirit of Western humanity looking at how low we've come, where we are and what kind of moral or ethical world we live in.... But as the genre developed, it was done to death; Hammett did it, Chandler did it, MacDonald did it. I think that

> there's really not much left to do. What else is Sam Spade going to do? He's static. What I'm doing with Easy, I'm making him get older and older, so the world around him is changing.... He's much more of an everyman in that case rather than an overman which is the way I see Sam Spade or Marlowe.[33]

An animus toward privilege, inequality and injustice is a common characteristic in the writings of all five novelists, who to an appreciable extent share the same liberal or reformist ideology. Macdonald, commenting on the genesis of *The Galton Case*, the most autobiographical of his detective novels, writes with considerable bitterness of 'the vision of the world which my adult imagination inherited from my childhood':

> It was a world profoundly divided, between the rich and the poor, the upright and the downcast, the sheep and the goats. We goats knew the moral pain inflicted not so much by poverty as by the doctrine, still common, that poverty is always deserved.[34]

The gulf between rich and poor is embodied thematically in *The Galton Case*, which uses the conventions of the mystery to keep the reader (and the detective) in suspense until the final pages: whether the claimant to the Galton estate, miraculously raised from poverty to privilege, is fraudulent or genuine. A similar complex plot reveals hidden connections between the seemingly separate worlds of wealth and poverty in Mosley's *White Butterfly* and *Black Betty* – almost as in *Bleak House*, only with an added racial dimension. Nearly all of Paretsky's novels present a parasitic, exploitative privileged class, protected by institutions which carefully guard their ascendancy, battening off the poor and vulnerable. Her female detective V.I. Warshawski frequently acts as tribune for the dispossessed: commenting on her first six novels, Paretsky sees her overall project as allied to that of Macdonald.

> Like Lew Archer before her she looks beyond the surface to 'the far side of the dollar', the side where power and money corrupt people into making criminal decisions to preserve their positions.[35]

Chandler is far less systematic in his social criticism and distrusted thematic interpretations of his detective fiction:

> So now there are guys talking about prose and other guys telling me I have a social conscience. P. Marlowe has as much social conscience

> as a horse. He has a personal conscience, which is an entirely different matter.

Yet in another letter he imputes a 'high moral content' to *The Big Sleep*, characterizing its theme as 'the struggle of all fundamentally honest men to make a decent living in a corrupt society'.[36] Explicit statements in Chandler's novels linking the accumulation of wealth with disease and corruption range from Marlowe's laconic 'The hell with the rich. They make me sick' in *The Big Sleep*, to the extended dialogue between Marlowe and Bernie Ohls in *The Long Goodbye*: 'Crime isn't a disease, it's a symptom. Cops are like a doctor that gives you aspirin for a brain tumor, except that a cop would rather cure it with a blackjack.... Organized crime is just the dirty side of the sharp dollar.'[37] The action of the novels and much of the individual imagery serve to reinforce the theme of wealth as disease: the opening scene of *The Big Sleep*, with the moribund General Sternwood in his stifling hothouse, the transformation of Velma into the wealthy Mrs Grayle in *Farewell, My Lovely*, the intricate pattern of deceit in *The Long Goodbye*. Toward the end of 'The Simple Art of Murder', Chandler generalises about the world depicted in his novels and those of Hammett:

> The realist in murder writes of a world in which gangsters can rule nations and almost rule cities ... a world where a judge with a cellar full of bootleg liquor can send a man to jail for having a pint in his pocket, where the mayor of your town may have condoned murder as an instrument of money-making.[38]

In Hammett's novels, the element of social criticism is more problematical because of the apparent amorality of his detective heroes who, unlike Marlowe, tend to share the values of the world they inhabit. Poisonville in *Red Harvest* is a nightmare vision of urban corruption, offering no possibility of redemption or relief (the novel's one sympathetic character, Dinah Brand, is thoroughly mercenary and unscrupulous, and is the seventeenth in the catalogue of murder victims). Hammett's Marxist beliefs, as Porter has suggested, may well be reflected in the depiction of 'an unregulated industrial capitalism, which acknowledges no limits to the pursuit of private wealth'. As studies of greed, betrayal and complicity in a world where no one is to be trusted, Hammett's novels are more pessimistic than those of Mosley, Paretsky and Chandler. James Ellroy, who considers the 'dark view' of *Red Harvest* similar to that in his own works, contrasts the novels of Chandler, which he accuses of being sentimental in their reformist 'liberalism', with those of Hammett: 'Hammett wrote the man he was afraid he was, where Chandler wrote the man he wanted to be.' [39]

Hammett and Chandler established the convention of the hard-boiled private eye, without family ties, emotional commitments, or a settled place in society, a perpetual outsider deeply distrustful of all institutionalized authority.

> The whole point is that the detective exists complete and entire and unchanged by anything that happens, that he is, as detective, outside the story and above it, and always will be. That is why he never gets the girl, never marries, never really has any private life, except insofar as he must eat and sleep and have a place to keep his clothes. His moral and intellectual force is that he gets nothing but his fee, for which he will if he can protect the innocent, guard the helpless and destroy the wicked, and the fact that he must do this while earning a living in a corrupt world is what makes him stand out.[40]

In most respects, Chandler's account fits both his own detective and Hammett's (indeed, it fits Holmes and Poirot nearly as well): though Sam Spade and the Continental Op are not idealized figures selflessly protecting the innocent, they are, like Marlowe, loners, instinctive celibates, who jealously guard their independence. Macdonald modifies Chandler's model only slightly, giving Archer a broken marriage in his background and making him brood about his loneliness, while lending a sympathetic ear to the emotional problems of his clients. Some recent writers in the hard-boiled school have imitated Chandler very closely: Michael Connolly, for example, makes his hero Harry Bosch an updated Marlowe with a badge:

> I'm fascinated by the idea of having an outsider in an insider's job. That's what I wanted Harry to be. I think the modern-day Philip Marlowe wouldn't be a private eye in today's L.A. He would be a cop, but he'd still have the kind of code Philip Marlowe had. He'd be a loner within the system.[41]

James Ellroy, whose principal figures also tend to be 'loners', differs from all these writers in rejecting Chandler entirely as a model. Unlike

Connolly, Paretsky and Mosley, Ellroy has constructed his own unheroic heroes almost as anti-Marlowes:

> What I wanted was to create my antidote to the ... predictable private eye. ... He gets weepy over lost dogs and little kids, he hates authority, he hates big money. ... I wanted a real, repressed, violent, right-wing ... L. A. cop.[42]

Paretsky and Mosley, though working within the same tradition as Hammett and Chandler, take a consciously revisionist stance in their detective novels: each has created complex, flawed central figures torn by contradictory impulses and evolving throughout a series of novels. Easy Rawlins and V.I. Warshawski, like Marlowe, are 'outsiders', sceptical toward authority in all its forms, careful to preserve their autonomy and privacy. Though far from celibate, each has great difficulty sustaining a sexual relationship: in *White Butterfly*, Easy's marriage collapses because of his habit of secrecy and instinctual distrust, and Vic resents the clumsy attempts of lovers and male associates to exercise control over her and treat her in a patronizing way. But both characters, rather than being unattached and isolated, are deeply embedded in a network of social and familial relationships, dependent on close, mutually sustaining (though sometimes difficult) friendships. Mosley has commented on this aspect of his novels:

> The genre has to develop. That *noir* character, who used to be outside of our lives is now inside our lives. ... I want Easy to be like everybody else. ... He's a regular guy, maybe a little better, maybe a little stronger, maybe a little smarter but he's a regular guy. He has children, he has a wife, that's the way life is. I'm not being critical of Ross MacDonald, Chandler, or Hammett, because they had a different project than mine but in today's world to write about a guy who doesn't have any responsibility or a woman who doesn't have any responsibility and so therefore can just make her decisions unhampered is a fantasy.[43]

A consistent attitude toward the police and the institutions of the state ostensibly devoted to the preservation of law and order can be found throughout the *noir* tradition. In the classic detective novel from Conan Doyle to Christie and Sayers, the gifted amateur is often contrasted with plodding policemen or unimaginative bureaucrats. But in Hammett, Chandler and their successors, the police are generally presented as corrupt, brutal, agents of injustice. In *Red Harvest,* Police Chief Noonan is the head of one of the rival criminal gangs bringing anarchy to Poisonville; Chandler's novels have a full complement of bullies, bigots, sadists, bribe-takers, even murderers among their policemen (along with the occasional honest cop like Bernie Ohls). The private eye in Chandler and Hammett not only distrusts authority, he is resented and disliked by those in authority. In Paretsky and Mosley, this mutual suspicion is reinforced by the issues of race, class and gender with which their novels are centrally concerned. The mutual hostility of Vic Warshawski and the police (most of whom, unlike the police in Chandler, are presented as honest but narrow-minded, as victims of false consciousness) comes largely from her position as a woman in a patriarchal society. The police in Paretsky's novels, as in Mosley's novels, are seen as one of many interlocking agencies upholding the values of a society riddled with corruption, where money rules and those without access to the instruments of power are ignored and ruthlessly exploited. As Paretsky says in an interview, 'My books give me the chance to nail the bad guys who ordinarily get away with it in the courts.'[44] To the police and the FBI in *White Butterfly* and *A Red Death,* all black people are potential criminals, and yet Easy finds himself in a painfully ambivalent relationship of complicity with the representatives of state authority in the complex working out of the plot. In Mosley's novels, which comprise a history of black life in Los Angeles over several decades beginning in 1948, the characters, faced with 'the racism of the world around them and the limitations in themselves that they were carrying with them', are beset by divided loyalties and exposed to the pressures of a society which simultaneously tempts and excludes them.[45]

The novels of Mosley illustrate the capacity of detective fiction to deal with serious issues and render a social milieu in convincing detail, while observing generic conventions. A recurrent preoccupation in Mosley's novels, as in those of Macdonald and Paretsky, is the American dream of success.

> It's not that you don't make it, it's that you realize that getting there was not what you thought it would be. And that really comes out of the *noir* genre, that whole idea of you can't really get what you want.[46]

Mosley's observations suggest that detective fiction in the hard-boiled tradition shares a number of ideological assumptions and thematic concerns with American mainstream fiction from Fitzgerald and Hemingway to Pynchon and Updike. As Chandler says in a letter, explaining why Marlowe and those like him must always be poor:

> There is absolutely no way for a man of this age to acquire a decent affluence in life without to some degree corrupting himself, without accepting the cold, clear fact that success is always and everywhere a racket.

The code of the private eye in Chandler and his successors is very close to the Hemingway code, with the significant difference that the detective hero, rather than making a separate peace, seeks to illuminate a 'hidden truth', of service to others, in the mean streets down which he (or in Paretsky's novels, she) must go.[47]

Notes

1 Franco Moretti, *Signs Taken for Wonders: Essays in the Sociology of Literary Forms*, trans Susan Fischer, David Forgacs and David Miller (London and New York: Verso, 1988), p. 138.

2 Moretti, *Signs Taken for Wonders*, p. 143; Dennis Porter, *The Pursuit of Crime: Art and Ideology in Detective Fiction* (New Haven and London: Yale University Press, 1981), p. 125. Both Porter and Marcus Klein argue that the American private eye novel of Hammett, Chandler and their successors is no exception in this respect. See Marcus Klein, *Easterns, Westerns and Private Eyes* (Madison: University of Wisconsin Press, 1994), p. 190: 'The vision is conservative, no matter Hammett's own subsequent involvement in left-wing politics. It is conservative in the classical and absolute sense of acceptance of depravity, while the style of the envisioning is weary, cool, knowing, and aloof.'

3 David I. Grossvogel, 'Agatha Christie: Containment of the Unknown', in Glenn W. Most and William W. Stowe (eds), *The Poetics of Murder: Detective Fiction and Literary Theory* (New York and London: Harcourt Brace Jovanovich, 1983), pp. 254–65. Christie's novels, Grossvogel argues, present 'a tidy world' in which 'law, order, and property are secure' (p. 265). Alison Light cites several instances of this conventional view of Christie in *Forever England: Femininity, Literature and Conservatism between the Wars* (London and New York: Routledge, 1991), pp. 62–4 and 234–5.

4 'The Simple Art of Murder', in Raymond Chandler, *Later Novels and Other Writings* (New York: Library of America, 1995), pp. 987 and 989; Chandler's dedication quoted in Ross Macdonald, *On Crime Writing* (Santa Barbara: Capra Press, 1973), p. 15.

5 Light, *Forever England*, pp. 88, 94 and 98.

6 W.H. Auden, 'The Guilty Vicarage', *The Dyer's Hand and Other Essays* (New York: Vintage Books, 1968), pp. 146–7 and 157–8. For an interesting commentary on guilt and anxiety in *The Murder of Roger Ackroyd*, a novel in

which the author 'cast doubt on the very conventions of narrative fiction', see Stephen Knight, *Form and Ideology in Crime Fiction* (London: Macmillan, 1980), pp. 112–15.

7 Nicholas Blake [C. Day Lewis], 'The Detective Story – Why?', in Howard Haycraft (ed.), *The Art of the Mystery Story* (New York: Grosset & Dunlap, 1946), p. 400.

8 John G. Cawelti, *Adventure, Mystery and Romance: Formula Stories in Art and Popular Culture* (Chicago and London: University of Chicago Press, 1976), pp. 95–6 and 106–7. Auden contrasts the treatment of guilt in *The Trial* and *Crime and Punishment* with that in detective fiction: 'The identification of fantasy is always an attempt to avoid one's own suffering; the identification of art is a sharing in the suffering of another.' ('The Guilty Vicarage', p. 158).

9 Geoffrey Hartman, 'Literature High and Low: The Case of the Mystery Story', in Most and Stowe (eds), *The Poetics of Murder*, p. 217; Ross Macdonald, 'The Writer as Detective Hero', in *On Crime Writing* , p. 11. Cf. Light, *Forever England*, p. 100: 'It is this dynamic which is embodied in the detective stories, the desire to stay snugly within known limits and the necessary urge to upset that tranquillity and test those boundaries.'

10 Richard Alewyn, 'The Origin of the Detective Novel', in Most and Stowe (eds), *The Poetics of Murder*, p. 69; Ronald Knox, 'A Detective Story Decalogue', in Haycraft (ed.), *The Art of the Mystery Story*, p. 196. Cf. R. Austin Freeman, 'The Art of the Detective Story', ibid., p. 15: 'The failure of the reader to perceive the evidential value of facts is the foundation on which detective fiction is built.'

11 Frank Kermode, 'Novel and Narrative', in Most and Stowe (eds), *The Poetics of Murder*, p. 184; Haycraft (ed.), *The Art of the Mystery Story*, pp. 187–99 and 273–86. Kermode sees the detective story as exemplifying 'the overdevelopment of one element of narrative at the expense of others: it is possible to tell a story in such a way that the principal object of the reader is to discover by an interpretation of clues, the answer to a problem posed at the outset' (p. 179).

12 R. Austin Freeman, in Haycraft (ed.), *The Art of the Mystery Story*, p. 11; Roger Caillois, 'The Detective Story as Game', in Most and Stowe (eds), *The Poetics of Murder*, p. 10; Auden, 'The Guilty Vicarage', p. 151.

13 Sayers, in Haycraft (ed.), *The Art of the Mystery Story*, p. 101. In another essay on detective fiction Sayers writes, 'To make an artistic unity it is, I feel, essential that the plot should derive from the setting, and that both should form part of the theme.' (ibid., p. 217)

14 Interview with Minette Walters, *The Armchair Detective*, 27 (1994), 185; *Selected Letters of Raymond Chandler*, ed. Frank MacShane (London: Macmillan, 1983), p. 135.

15 Light, *Forever England*, pp. 91–2. Cf. 'The Philosophy of Composition' in Edgar Allan Poe, *Selected Writings* (Harmondsworth: Penguin, 1978), p. 480.

16 Ross Macdonald, *Self-Portrait: Ceaselessly Into the Past* (Santa Barbara: Capra Press, 1981), pp. 88 and 93; *On Crime Writing*, p. 22. Macdonald contrasts his own novels with those of Chandler, from whom he 'learned a great deal': 'Chandler described a good plot as one that made for good scenes, as if the parts were greater than the whole' (*On Crime Writing*, p. 22). Chandler's plots, like Hammett's, tend to be episodic and crowded with

incidents. As he says in a letter, 'You never quite know where your story is until you have written the first draft of it.... What seems to be alive in it is what belongs in the story' (*Selected Letters*, pp. 87–8).

17 Auden, 'The Guilty Vicarage', p. 158; Tzvetan Todorov, *The Poetics of Prose*, trans. Richard Howard (Oxford: Basil Blackwell, 1977), p. 43. See also Cawelti, *Adventure, Mystery and Romance*, pp. 9–15.

18 Edmund Wilson, 'Who Cares Who Killed Roger Ackroyd?', in Haycraft (ed.), *The Art of the Mystery Story*, pp. 390–7; Marjorie Nicolson, 'The Professor and the Detective', ibid., pp. 110–27; Joseph Wood Krutch, 'Only a Detective Story', ibid., pp. 178–85.

19 Sayers, in Haycraft (ed.), *The Art of the Mystery Story*, pp. 101, 102 and 109.

20 Chandler, *Selected Letters*, pp. 134 and 181; Chandler, *Later Novels*, p. 986. Elsewhere Chandler says in a letter to Erle Stanley Gardiner, a friend and best-selling detective novelist, 'When a book, any sort of book, reaches a certain intensity of artistic performance, it becomes literature.' (*Selected Letters*, p. 69)

21 Sayers, in Haycraft (ed.), *The Art of the Mystery Story*, p. 209. Sayers attributes this view to S.S. Van Dine (see ibid., pp. 33–42). Chandler is unequivocal in his view of 'the classic detective story': 'They do not really come off intellectually as problems, and they do not come off artistically as fiction. They are too contrived, and too little aware of what goes on in the world.' (*Later Novels*, p. 985)

22 Sayers, in Haycraft (ed.), *The Art of the Mystery Story*, pp. 209 and 218. The quotations are from an interesting essay, written after the publication of *Gaudy Night* in 1937, which traces the evolution of the Wimsey novels, as they more and more become concerned with serious themes.

23 *Selected Letters*, pp. 170 and 180; *Raymond Chandler Speaking*, ed. Dorothy Gardiner and Kathrine Sorley Walker (London: Hamish Hamilton, 1962), p. 57. In another letter, Chandler writes, 'I concentrated on the detective story because it was a popular form and I thought the right and lucky man might finally make it into literature.' (*Selected Letters*, p. 459)

24 Chandler, *Later Novels*, pp. 988 and 989–90.

25 Chandler, *Later Novels*, p. 989; *Selected Letters*, p. 4; Macdonald, *On Crime Writing*, p. 15. Chandler makes the comparison to Hemingway in 'The Simple Art of Murder', *Later Novels*, p. 988. Critics discussing the style of Chandler and Hammett as conscious fabrication include Porter, *The Pursuit of Crime*, pp. 130–45; Knight, *Form and Ideology*, pp. 139–49; and Fredric R. Jameson, 'On Raymond Chandler', Most and Stowe (eds), *The Poetics of Murder*, pp. 122–48.

26 Chandler, *Selected Letters*, p. 115.

27 Chandler, *Selected Letters*, p. 174; *Later Novels*, pp. 989–90.

28 Macdonald, *On Crime Fiction*, pp. 15, 16 and 19; Steven Marcus, 'Dashiell Hammett', in Most and Stowe (eds), *The Poetics of Murder*, p. 208.

29 Chandler, *Selected Letters*, p. 173.

30 'The other side of those mean streets', interview with Walter Mosley, *The Armchair Detective*, 26 (1993), 12–13.

31 Macdonald, *On Crime Fiction*, p. 24; *Self-Portrait*, p. 43.

32 On the implicit sexism in 'the private investigator's code' in Hammett and Chandler, see Porter, *The Pursuit of Crime*, pp. 183–6. On 'the tensions

between the demands of the detective novel and the feminist ideology' of Paretsky's novels, see Kathleen Gregory Klein, *The Woman Detective: Gender and Genre* (Urbana and Chicago: University of Illinois Press, 1988), pp. 200–16.

33 Interview with Walter Mosley, *The Armchair Detective,* 25 (1993), 12. Macdonald similarly sought to create in Archer 'a hero who sometimes verges on being an anti-hero', reacting against what he considered the excessive idealization in Chandler's hero: see Macdonald, *On Crime Writing,* pp. 23–4.

34 Macdonald, *On Crime Writing,* p. 32.

35 Paretsky's comment occurs in *Twentieth-Century Crime and Mystery Writers,* ed. Lesley Henderson, 3rd edn, (Chicago and London: St James Press, 1991), p. 829. In interviews, Paretsky has emphasized the political dimension of her novels: 'V.I. shares my politics.... V.I. has given me a voice, given me the courage to say a lot of things I wouldn't say in my own voice': *Books,* 6 (May/June 1992), 4.

36 Chandler, *Selected Letters,* pp. 43 and 197.

37 Raymond Chandler, *Stories and Early Novels* (New York: Library of America, 1995), p. 636; *Later Novels,* p. 712.

38 Chandler, *Later Novels,* p. 991.

39 Porter, *The Pursuit of Crime,* p. 197; interview with James Ellroy, *The Armchair Detective,* 28 (1998), 240–1.

40 Frank MacShane, *The Life of Raymond Chandler* (London: Jonathan Cape, 1976), p. 70.

41 Interview with Michael Connolly, *The Armchair Detective,* 28 (1995), 400.

42 Interview with Ellroy in John Williams, *Into the Badlands: Travels through Urban America* (London: Flamingo, 1993), p. 90.

43 *The Armchair Detective,* 26 (1993), 14. On the theme of women's friendship in Paretsky, see the chapter in this volume by Margaret Kinsman, 'A Band of Sisters'.

44 Interview in the *Guardian,* 25 August 1987, p. 8, quoted in Sally R. Munt, *Murder by the Book? Feminism and the Crime Novel* (London and New York: Routledge, 1994), p. 42.

45 *The Armchair Detective,* 26 (1993), 12.

46 Ibid.

47 Chandler, *Later Novels,* pp. 991–2; *Selected Letters,* p. 197.

10

Authority, Social Anxiety and the Body in Crime Fiction: Patricia Cornwell's *Unnatural Exposure*

Peter Messent

How do we explain the attraction crime literature holds for its readers? Any answer to this question is going to be necessarily tentative and speculative given the problems both in identifying different readerly communities and the responses (both individual and shared) that compose them; given, too, the different social and historical circumstances in which such texts have been produced and received. Contemporary American crime fiction, traced back via Chandler and Hammett, has predominantly urban roots. The subject of David Stewart's recent article in *American Literary History* is an even earlier period, pre-Civil War urban crime literature and spectator journalism in particular.[1] Nonetheless, his explanatory model and his observations on the related economies of production, pleasure and desire offer a helpful prompt for thinking about contemporary crime fiction and exploring the nature of its appeal.

Stewart sees the 'relish' (p. 681) with which mid-nineteenth-century crime literature was read in terms of the ambivalent attitudes held about criminality and the dominant social order of the time. He identifies the popular appeal of such writing in the way it 'eroticized' urban experience, providing a necessary and thrilling release from the disciplinary procedures of capitalism, from the 'laws and behavioural practices' (p. 689) that sustained an increasingly regimented social order. The 'exhilaration' (p. 688) associated with criminal danger and the 'dark underside' (p. 689) of city life, then, stood as a direct and exciting contrast to a daily experience 'that was, for the majority of city dwellers, constraining, confining, and mind-numbingly dull' (p. 684). But crime writing also engaged quite opposite emotions, feeding on a popular *fear* of crime in the threat to the reader's own security contained in such 'narratives of violation' (p. 682). The desire for transgressive excitement and the need for safety are then ambivalently weighed against each another.

Another crucial and analogous tension that Stewart identifies in this literature is the way in which the qualities of rational order and epistemological control which crime fiction associates with the representative of the law (detective or police) are undermined and opposed by those descriptions of the spilling of blood, sexual abuses and physical suffering which characterize the genre. One newspaper story evokes the corpse, 'the victim of bodily opening and exposure', in gory detail.[2] As he describes this, Stewart asks,

> what, if any, of the information is in fact evidence? ... And more disturbing: Why is it important to differentiate among drops, clots, and pools [of blood]? What is the depth of a razor slash evidence of? ... Prurient excess would seem to undermine the strict task of productive looking. More to the point, productive looking seems to lead all too irresistibly to prurient excess. (p. 695)

How are we to understand such 'nonproductive desires associated with violence' (p. 694) and the readers' fascination and abhorrence? Why is so much time spent giving details that are irrelevant to the scientific business of detection? Do such bloody excesses again signify for the reader some psychological compensation for a 'behavioural regime' (p. 696) that constricts the day-to-day social practice of human behaviour into its most productive channels?[3] I shall return to some of these questions later in this chapter. For the moment it is enough to note the complex and divided attitudes toward law and violence, day-to-day social practice and its bloody transgression, which mark both the practice of crime literature and the readerly responses it stimulates.

Stewart's identification of some of the ambivalences and paradoxes present in early urban crime writing and the responses to it offers confirmation for my own awareness of the contradictions in the way *contemporary* crime fiction is assessed. Stewart's identification of a tension within crime writing between forms of rational control and an escape beyond its borders points to a related paradox that becomes apparent in different critical accounts of the genre itself. Thus on the one hand we find an insistence on the conservative nature of the form; on the other, an interest in what Catherine Nickerson calls the 'explosive cultural material' it contains.[4]

Dennis Porter argues the former case by noting how the genre reinforces the dominant social order, for 'in a detective story ... the law itself is never put on trial.'[5] Franco Moretti puts this even more bluntly, implicating the detective her- or himself in this process: 'the detective is the figure of the state in the guise of "night watchman", who limits himself to assuring respect for laws'.[6] But if the detective is constituted as law-bringer, and if the very form of such fiction operates to deny social change, how do the 'disturbing forces'[7] which break loose within such boundaries work to any more than shallow effect? Nickerson implies that they work in highly significant ways, claiming that the detective novel (and its visual equivalents) are

> *deeply* enmeshed with most of the thornier problems of the Victorian, modern, and postmodern eras, including gender roles and privileges, racial prejudice and the formation of racial consciousness, the significance and morality of wealth and capital, and the conflicting demands of privacy and social control.[8]

Thus, for her, the genre 'represent[s] in a generally realistic style the most anxiety-producing issues and narratives of a culture'.

A recent crime novel, Patricia Cornwell's *Unnatural Exposure* (1997) demonstrates the relevance of these introductory remarks to current fictional practice. I shall discuss the larger contradiction just identified (the exposure of thorny cultural problems within a conservative generic frame) and – recalling the razor slash and the clots of blood to which Stewart refers – the exploration of bodily anxieties as they are represented in Cornwell's text. If a novelist such as Walter Mosley raises potentially explosive social issues with the representations of race and masculinity in Los Angeles urban space in his novels, then Cornwell works at the more conservative end of the detective fiction spectrum. She does, however, address contemporary gender issues in her fiction, and these have recently become increasingly prominent in feminist crime fiction.[9] Evelyne Keitel's comments on the images of self-sufficiency portrayed in such texts suggest how such fiction allows an identification with female agency and provides a version of the power of the subject in a period when the very possibility of such an identification (with either male *or* female agency) is facing considerable challenge: 'Feminist mysteries ... compensate [both] for women's lack of social space in which to act [and for] the precariousness of the post-structuralist subject position.'[10]

Unnatural Exposure is the latest novel in Cornwell's series featuring Kay Scarpetta, the Chief Medical Examiner for Virginia, based in Richmond. Again and again, she comes up against men and women who do not expect a woman to be in a position of authority, and Cornwell's novel plays on the gap between such expectations and Scarpetta's challenge to them:

> 'They said the medical examiner would get here, and for me to watch for him', he said to me.
> 'Well, that would be me', I blandly replied.
> 'Oh yes, ma'am. I didn't mean anything...' His voice trailed off.[11]

Scarpetta's intelligence, expertise, tough assertiveness and success in her ('unfeminine') professional area – which work to subvert conventional gender assumptions – are all emphasised in the novel.[12]

Cornwell also addresses the problems confronting a woman writer working with what is recognised as a 'masculine genre'.[13] Bodily violations and exposure play a considerable part in Cornwell's fiction (as one would expect with a pathologist as main protagonist), and the appearance of the first dismembered body on a landfill site on the outskirts of Richmond would initially seem to promise the connection between urban anxieties and female victimization on which the crime novel commonly relies. James Ellroy's *The Black Dahlia* (1987) is an obvious example, with the early discovery in a vacant Los Angeles lot of the sexually mutilated body of an attractive young woman and the eventual connection of this killing with the dynamics of the city's property market in the pre-war period. A similar pattern appears to emerge in *Unnatural Exposure*, with the dismembered remains of a female murder victim symbiotically linked to the remnants of urban consumption, found among other 'decomposing trash' (p. 21) on the landfill site.

At first glance, too, Cornwell (like Ellroy) seems to be concerned with serial crime here, the 'illegibility' (meaninglessness, apparent motivelessness) of which can be seen to indicate the extremes of affectlessness and narrative incoherence that mark the postmodern social order. Thus according to Barry Taylor,

> As...the sign of a threatening randomness, of a disappearance of meaningful inter-subjective structures, of demotivated action, of the collapse of authoritative models of explanation and interpretation, of 'classical' narrative, and of the disappearance of the subject, serial murder appears to be situated at the gathering point of a number of themes which are central to a variety of discourses on the postmodern.[14]

In the case of *Unnatural Exposure*, however, such connections prove something of a false trail. The narrative swerves from its starting premise, an account of serial murder, as Scarpetta deduces that this death is a one-off 'copycat' version of prior murders, and as the case turns out to involve a close professional and personal connection between the criminal and the investigator quite different from the urban pathologies with which the genre is normally associated. For it is Phyllis Crowder, a microbiologist and medical colleague of Scarpetta, who turns out to be the murderer. Her motives include both being passed over for promotion in her Richmond job and resentment at Scarpetta's own success. Though two of the four Crowder victims are female, neither is young or attractive,[15] and considerable play throughout the text is made on the gap between the assumption that the murderer is male and the fact that she is in reality female. To a certain extent, Cornwell undermines 'the unshakable hierarchical pairs of mind/body and male/female' that traditionally structure the crime fiction genre (at least in its hard-boiled manifestations).[16]

Cornwell also gives a sympathetic representation of lesbianism, and what it means to work as a lesbian in a predominantly male and physically active profession, in her representation of Scarpetta's niece Lucy, a member of the HRT – the FBI's Hostage Rescue Team. The displacement of Cornwell's own lesbian identity away from the main protagonist to her niece may suggest that an interrogation of 'existing gender definitions and concepts of sexual politics'[17] is not the main item on Cornwell's agenda – although, of course, it may be the market that motivates her to consign lesbianism to a secondary textual level. But it is the stress on Scarpetta's own heterosexual identity that begins to confirm, even in the realm of gender politics, the more conservative aspects of the book. For her anxieties about the relationship between the private and professional demands in her life, centring on Benton Wesley, the FBI executive whose on–off relationship with Scarpetta has formed a continuing strand to the series, are apparently finally resolved in this novel. That this resolution takes the form of Scarpetta's full romantic commitment – the novel ends with the words '"I love you, Benton," I said.' – is highly significant.

One might claim more generally that any interrogation Cornwell makes of the established social order is outweighed by her defence of it. First, we might note the status of this text as a type of police procedural and thus part of a recent shift of emphasis away from private-eye crime fiction, and from the rule-bending individualism and extra-systemic freedom manifested by its detective protagonists.[18] Scarpetta has both state and national police powers at her disposal (she is Consulting Forensic Pathologist for the FBI) as she pursues her investigative work. Joyce Carol Oates's description of the irrelevance of the private eye in contemporary criminal investigations is pertinent here: 'private detectives are rarely involved in authentic crime cases, and would have no access, in contemporary times, to the findings of forensics experts.'[19] Cornwell's novels rely on just such 'scientific detection' (of which DNA-tracing is the most obvious example) and the ready access to local and national surveillance services that Scarpetta's official position brings her. Thus, as is usual in the police procedural, Scarpetta's actions ultimately and necessarily serve to preserve the status quo.

Moreover, if a debate about the nature of sexual politics in America is, as I have suggested, carried out within this generic frame, such social debate only operates within a narrow field, one bound by 'the privileges of wealth and whiteness'.[20] Despite references to Scarpetta's Italian-American background, and the use of Tangier, a small island in Chesapeake Bay, and the archaic dialect of its inhabitants as a textual element, there is little real concern here with poverty, or with ethnic, racial or class difference (what Nickerson calls the 'thornier problems' of contemporary American life). It is true that there is an ongoing concern with a recent American political problem, the budget impasse in Washington that brings federal agencies to a virtual standstill, and both the solution and containment of Crowder's crime are threatened, but this concern is marginal. Scarpetta lives in a wealthy suburb, and although one of Scarpetta's clerks is called Cleta (African American?), a Coast Guard chief is called Martinez (but with the first name Ron), and an Army Colonel is called Fujitsubo, such indications of different socio-economic worlds and of multi-cultural difference count finally for very little. Indeed, there is an odd and perhaps revealing passage when Scarpetta checks out the various chat rooms available on the Internet (for the criminal, Crowder, teasingly communicates with Scarpetta on e-mail and computer chat room) to comment that:

> There was truly something for everyone, places for flirts, singles, gays, lesbians, Native Americans, African Americans, and for evil. People who preferred bondage, sadomasochism, group sex, bestiality, incest were welcome to find each other and exchange pornographic art. (p. 210)

There is just a hint here, in the slippages of Scarpetta's language, and despite the sympathetic representation of lesbianism in the novel, that the further away from a heterosexual white American norm one moves, the closer one gets to 'evil'. Equally, the move from particular social types to metaphysical category (evil) and from various forms of unconventional sexual practice to trans-cultural taboo (incest) suggests something of the conservative novelistic imagination at work here.

Such tendencies are similarly implicit in the description of criminal activity the text portrays. Anxieties about the violent underbelly of urban American life might be raised in the initial discovery of the severed female corpse, and rather different anxieties might then be triggered about the ability of a single crazed individual to hold a nation or the world to ransom through a form of germ warfare. (We should also note that it is finally and fortuitously the fact that the virus Crowder creates is non-replicating, not Scarpetta's detective work, that prevents wide-spread disaster.) For the eruption of crime in this novel is signalled by the 'eruptions' (p. 55) and 'fulminating pustules' (p. 330) that mark the bodies of its victims – Crowder has unleashed a mutant version of the smallpox virus on her chosen victims. But Cornwell's consignment here of crime to the category of pure evil, dependent on the purely personal motivation and disturbed psychological condition of the single criminal figure, removes it effectively from the world of social cause and effect. At the start of the book, an administrator at the Dublin coroner's office tells Scarpetta that 'American violence is so exotic to us'. She replies, 'That's rather much like calling a plague *exotic*.' (p. 7) This, together with the book's epigraph from the Book of Revelation ('seven angels [with] ... seven vials full of the seven last plagues') and the possible release of a variant of a smallpox plague as its main plot-line, come together to suggest that *all* criminal violence is plague-like, part of a Manichean battle between good and evil that is inevitably waged in human societies, taking an extreme form in America with its prevalence of serial murders ('the Bundys, the Dahmers', p. 333) and its conspicuous acts of random and widespread violence ('the Unabomber', p. 143).

Such a way of representing crime is parallelled by Cornwell's constant use of exceptionalist discourse in the text. Thus the victims of the initial serial murderer are described as 'nothing but symbols of his private, evil credo' (p. 11), while Crowder, her identity unknown, is simply 'the monster' (p. 199). When she is then known, Crowder's eyes are described as 'like evil', and she is punished not by law but by a more immediate form of justice, by dying the way her victims did. The point is that although Cornwell is concerned with the various types and extreme forms of violent crime in America, her work does not fully engage 'the most anxiety-provoking issues and narratives' of American culture. For to consign crime to the world of pure evil and individual moral monstrosity is to isolate it from all economic, social or political causes and to explain it as a psychopathic and freakish exception to all that we know to be 'normal'. Cornwell's is a black-and-white world where evil is pure 'other', finally defeated partly just by luck – or perhaps by God's will? Scarpetta is after all a Catholic (p. 219) – and partly by the rational, analytic and commanding figure of the female detective working on behalf of the established social order.[21]

There is one further crucial issue at stake here, and this is where I return to the depth of those razor cuts about which Stewart is so concerned. For there is undoubtedly something deeply disturbing about a series of fictions with, in Sabine Vanacker's words, a 'woman hero [who] is a dealer in death, who aggressively "manhandles" the corpses of victims and gruesomely thrives off decaying and decomposing bodies'.[22] And the growing popularity of what we might call the forensic thriller – as evidenced in recent British television series such as *Silent Witness* and *McCallum* – suggests that such a focus on dead bodies and their dissection holds considerable interest for a popular audience. Thus if crime novels such as Cornwell's are to be identified with explosive cultural material, this may well primarily pertain to the representations of the body violations we find there.

It is the series of explicit references to the body infected by HIV and AIDS, however, which might seem to provide the more immediate engagement with current 'thorny' social problems in the text.[23] But Cornwell's depiction of Wingo, the gay man who works with Scarpetta and runs the Richmond morgue, and his status as HIV positive, only gives us more evidence of the covert forms of conservatism which mark this novel. At first glance, Cornwell's attitude toward AIDS would appear to be a highly sympathetic one. When Wingo first comes to Scarpetta to tell her he is HIV positive, she prefaces his revelations by saying,

> your life is no secret to me. I don't make judgements. I don't label. In my mind, there are only two categories of people in the world. Those who are good. And those who aren't. But I worry about you because your orientation places you at risk. (p. 64)

Scarpetta's liberal credentials on this matter are then amply reinforced in her empathy and support for him when he tells both of his condition and of his fear of AIDS: 'You will not go through this alone. You have me.... I will take care of you' (p. 65). But the fact that Wingo is the one member of Scarpetta's staff who is infected by the virus loosed by Crowder suggests a different reading. For in a novel where a reference to a divinely-inspired plague is introduced in the book's epigraph and where the notion of the criminal dissemination of a highly infectious disease with potentially world-wide epidemic consequences stands as the central plot-line, the fact that it is the HIV-positive Wingo who catches the smallpox variant and dies from it, is surely far from coincidental. The social and moral sub-text seems clear. The AIDS motif is a needless irrelevance in the novel unless it too is to be linked to the sense of criminal revulsion and moral condemnation associated with the other threatened 'epidemic' (p. 223). Thus, Cornwell's apparently liberal agenda proves on close inspection to be somewhat misleading. The codes and values of the so-called moral majority are those ultimately endorsed here.

I return, then, to the prevalence of mutilation, dissection, decay and death in Cornwell. For it is in its treatment of body horror that I wish to locate the symbolically disturbing core of her fictions. The concern with this subject in recent film criticism provides a useful entry point to my analysis here. In a 1995 article on the horror genre, Barbara Creed draws on Kristeva's work on the abject to explain the changing representation and symbolic significance of the body in this popular form. The abject is defined as 'everything that challenges the subject's identity as human'.[24] Blood, torn flesh, spilling entrails and dismemberments are seen as part of this challenge in their threat to the integrity of 'the clean and proper body' in transgressing the boundary between the human and the nonhuman. Thus 'the ultimate in abjection is the corpse.'[25]

To follow through this argument and to apply it to crime fiction would be to interpret that genre's shared interest in bodily violation in a number of different ways. Most straightforwardly, Creed identifies such abjection ('the body in profuse disarray')[26] as 'displace[d] anxiety' (p. 143) about the status of the subject; as symptomatic of 'an increasing sense of individual helplessness' (p. 129) and a fear of loss of individual authority and control in this (postmodern) world. Fredric Jameson characterizes postmodernism in similar terms (without realizing how literally his words might be applied) when he speaks of 'the alienation of the subject [being] displaced by the fragmentation of the subject'.[27] As Creed writes, 'images of the bleeding body... point symbolically to the fragile nature of the self, its lack of secure boundaries, the ease with which it might lose definition, fall apart, or bleed into nothingness' (p. 144). We can then go on to argue that in its violations of the clean and proper body, its denial of 'the sanctity of life' (p. 132), and in the upsetting of law and the normative routines of community life which it enacts, crime fiction indicates our considerable anxieties about the successful functioning of the contemporary social system as a whole.

According to Creed, then, as we read of the textual victim's bodily scarring or mutilation, we reconstruct a sense of our own 'body as clean, whole, impregnable, living, inviolate'. Such reading reaffirms 'a comforting but illusory sense of a unified, coherent, authentic body and self' (pp. 156-7). But an alternative response is equally possible: one which 'reminds [the reader] of the fragile nature of all limits and boundaries' (p. 157) and which fearfully 'shakes our sense of bourgeois identity' (p. 154) in its denial, via the representation of the victim and his or her abject body, of autonomy and of firm social belonging, of the protections generally associated with firm social hierarchies and established communities.[28] Or, to take this one stage further, such texts can, in their violent excess, offer a balance to the repressive aspects of this same bourgeois identity, provide a compensatory escape from, or an ironic response to, the restrictive parameters of 'law, rationality, logic' (p. 137).

Cornwell indirectly addresses such questions in her various descriptions of onlookers at crime scenes 'gawking' at the victims: 'Looking was too much of a temptation for most people to resist. The more gruesome the case, the more this was true.'[29] If Scarpetta first takes the reader to the scenes where violated bodies lie, and then in a sense compounds that violation in her role as pathologist,[30] why, then, does the reader want to 'gawk' at the spectacle presented, want to read about Scarpetta as, for instance, she defleshes and degreases the bones of a murder victim (p. 58) or as she cuts up sections of a lifeless heart, or pulls loosened face-skin 'forward over the eyes to expose the skull', in a way which becomes something like routine?[31] If there is an ironic response at work here, it may lie in the way that bloody excess and our mixed feelings of abhorrence for, and attraction to, such excess (Cornwell's books are, after all, very popular), are a symptom of our detached symbolic recognition of just how completely the social order has failed us, and leads us to interrogate sardonically the belief in law and logic on which our social system is grounded. To repeat, however, this is just one of a number of possible interpretive responses which stand one against the other in a necessarily unstable and ambiguous relationship.[32]

Can we say more about Scarpetta's own role as pathologist and detective in these fictions? Perhaps we should remind ourselves that detective fiction is at issue here and that the connections between horror film and crime fiction that have been drawn cannot stand too much pressure. If there is a similarity in their joint concern with images of blood and bodily disfiguration, Creed's argument concerning 'the horror film's attack on the symbolic order'[33] and the carnivalesque reversals occurring within the genre can only be applied in limited form to crime fiction, where the figure of the detective and the domain of Law retain a constant counterbalancing presence. But that leaves us with the question of what to make, in Cornwell's fiction, of a fictional protagonist who is responsible for the further openings and cuttings of the body at the same time as she responds to its criminal disfiguring, and one moreover who 'has radically crossed the limits of her gender role, with her choice of the most unsavoury and "unfeminine" of professions'.[34] That Scarpetta speaks the word of the law, and that her dealings with the materials of the abject – her scientific probings of the bodies of the dead – take place ultimately on a rational and scientific level, can be seen as reassurance in the face of the threatening and disruptive powers of the bodily realm.[35] Nonetheless, the obsession with bodily openings and the processes of its dissection represented is in itself problematical and unsettling.

One further element, however, enters the already complicated picture here, in Cornwell's twinning of criminal and pathologist–detective in this novel. In communicating with Scarpetta by e-mail and computer chat room (sending photos of her first victim via the screen name 'deadoc'), Crowder establishes her identity as Scarpetta's dark double, describing herself in the subscribers' directory with Scarpetta's own personal and professional details: ' "It's like deadoc's saying he's you," Lucy said' (p. 109). The motif of contamination raised early in the text with Scarpetta's memory of the James river flood, 'water poisoned pink with formalin seeping into the morgue and the parking lot in back' (p. 54), is evident here once more, as the same professional talents that Scarpetta possesses are turned, in Crowder's case, against the law. This twinning of the figures – one e-mail message runs '*death doctor death you are me*' (p. 280) – also raises questions about the precise nature of Scarpetta's professional interests in death, savaged bodies and bodily decay. Scarpetta's obsession with the abject and the fascinations and disturbances associated with that realm suggest a deep ambivalence about identity, the social order and their ability to sustain each other that runs through all of Cornwell's writing. And such statements as 'the dead have never bothered me. It is the living I fear' are symptomatic of one who is rather too deeply immersed in the province of abjection for comfort.[36] Stewart's earlier contrast between the world of productive work and contrasting forms of psycho-social release gets peculiarly muddled here (and again we should remember that contamination is a crucial metaphor in the text) as Scarpetta appears fascinated by the contemplation of death at the very same time as she is most productively efficient in her sustaining of that social order which battles against it. There are deep ambivalences contained in such a representation.[37]

Notes

1 David M. Stewart, 'Cultural Work, City Crime, Reading Pleasure', *American Literary History*, 9.4 (1997) 676–701. Page references are indicated in parentheses in the text. Stewart bases his essay on ideas of productivity and non-productivity and the symbolic exchanges that take place between these two spheres within a capitalist economy.
2 'Horrible and Mysterious Murder in Broadway', *Frank Leslie's Illustrated Newspaper* (2 August 1856).
3 Stewart answers affirmatively, drawing on Georges Bataille, *Eroticism: Death and Sensuality*, trans. Mary Dalwood (San Francisco: City Lights, 1986). In the newspaper story, the murdered man dies in an apartment which is both home and workplace (he is a porter). Thus his 'company's demands for regularity, respectability, and productivity' are symbolically jammed by the bloody body which 'successfully defaced productive company space' (Stewart, p. 695). Bataille argues for the erotic appeal of death as a counter to the world of work and 'productive efficiency' (*Eroticism*, p. 41). 'In the end', he contends, 'we resolutely desire that which imperils our life' (p. 86).
4 Catherine Nickerson, 'Murder as Social Criticism', *American Literary History*, 9.4 (1997) 756.
5 Dennis Porter, *The Pursuit of Crime: Art and Ideology in Detective Fiction* (New Haven: Yale University Press, 1981) p. 122.
6 Franco Moretti, *Signs Taken for Wonders: Essays in the Sociology of Literary Forms* (London: Verso, 1983), pp. 154–5. While Moretti is discussing the European tradition and the figure of the private eye in American hard-boiled detective fiction bears a more marginal relationship to the law, the argument remains essentially the same. See 'From Private Eye to Police Procedural: The Logic of Contemporary Crime Fiction', Introduction to Peter Messent (ed.), *Criminal Proceedings: The Contemporary American Crime Novel* (London: Pluto, 1997), pp. 1–21.
7 Moretti, *Signs Taken for Wonders*, p. 137.
8 Nickerson, 'Murder as Social Criticism', p. 744, my emphasis added.
9 I am thinking especially of Sue Grafton, Sara Paretsky and Cornwell. Grafton hardbacks now have initial print runs of 500000 copies.
10 Evelyne Keitel, 'The Woman's Private Eye View', *Amerikastudien/American Studies*, 39 (1994), 178–9.

11 Patricia Cornwell, *Unnatural Exposure* (London: Little, Brown, 1997), p. 21. See also pp. 168 and 191–3. Page references are indicated in parentheses in the text.

12 For suggestive comments on Cornwell's representation of the successful but solitary woman in a male professional hierarchy and her interrogation through Scarpetta of 'the masculine ideal of objectified knowledge', see Sabine Vanacker, 'V.I. Warshawski, Kinsey Millhone and Kay Scarpetta: Creating a Feminist Detective Hero', in Messent (ed.), *Criminal Proceedings*, pp. 77 and 81.

13 See Nickerson, 'Murder as Social Criticism', pp. 750–1.

14 See 'The Violence of the Event: Hannibal Lecter in the Lyotardian Sublime', in Steven Earnshaw (ed.), *Postmodern Surroundings* (Amsterdam and Atlanta, Georgia: Rodopi, 1994), p. 217.

15 If the torso of the female victim is described as 'a hideous stub' with 'maggots teeming in the genital area' (p. 24), the sexual associations raised here are once again a false trail, acting as prelude to a narrative that resists the normal generic gender stereotyping.

16 Nickerson, 'Murder as Social Criticism', p. 751. Nickerson is here condensing Kathleen Gregory Klein's arguments from an essay in Glenwood Irons (ed.), *Feminism in Women's Detective Fiction* (University of Toronto Press, 1995).

17 See Paulina Palmer, 'The Lesbian Thriller: Transgressive Investigations', in Messent (ed.), *Criminal Proceedings*, p. 88.

18 Though in fact the actions of even this protagonist remain subordinated to larger forms of social monitoring and control. See 'Introduction', *Criminal Proceedings*, pp. 1–21.

19 Joyce Carol Oates, 'The Simple Art of Murder', *The New York Review of Books*, 45.20 (21 December 1995), 34–5.

20 Nickerson, 'Murder as Social Criticism', p. 750, paraphrasing from Bobbie Ann Mason, 'Nancy Drew: Once and Future Prom Queen', in Irons (ed.), *Feminism in Women's Detective Fiction*.

21 I recap the argument here from the 'Introduction' to *Criminal Proceedings*, p. 16.

22 Vanacker, 'V.I. Warshawski, Kinsey Millhone and Kay Scarpetta', p. 66.

23 Recent popular culture, in the form of vampire movies, has symbolically foregrounded the public concern with AIDS in its dominant motif of infection via the transfer of bodily fluids.

24 Barbara Creed, 'Horror and the Carnivalesque: The Body-monstrous', in Leslie Devereux and Roger Hillman (eds), *Fields of Vision: Essays in Film Studies, Visual Anthropology, and Photography* (Berkeley: University of California Press, 1995) p. 149. Page references are indicated in parentheses in the text. Julia Kristeva writes on the abject in *Powers of Horror: An Essay on Abjection* (New York: Columbia University Press, 1982).

25 Creed quotes here from her own 'Horror and the Monstrous-Feminine: An Imaginary Abjection', *Screen*, 27.1 (January/February, 1986), 46–7.

26 The phrase is taken from Pete Boss, 'Vile Bodies and Bad Medicine', *Screen*, 27.1 (January/February, 1986), 15.

27 Fredric Jameson, 'Postmodernism, or the Cultural Logic of Late Capitalism', *New Left Review*, 146 (July/August 1984) 79–80. See also Boss, 'Vile Bodies and Bad Medicine', pp. 22–3.

28 Boss identifies one area in which such fears operate in terms of our 'powerfully ambivalent sense of the institution of modern medicine'. If the technology and power of medicine have improved our condition, nonetheless 'despite the immaculate order of the hospital, its brilliance and asepticism ... it remains a sanctuary of contemporary terror'. This terror lies in the way that the subject becomes 'defenceless matter' in this institution, 'one's own body rendered alien, regulated, labelled, categorised, rearranged, manipulated, scrutinised and dissected'. See 'Vile Bodies and Bad Medicine', pp. 19–20.
29 Cornwell, *Unnatural Exposure*, pp. 23 and 21.
30 See the discussion in Vanacker, 'V.I. Warshawski, Kinsey Millhone and Kay Scarpetta', p. 83: 'She performs tasks that fly in the face of nature, opening up what must remain closed.'
31 Patricia Cornwell, *Cause of Death* (1996; London: Warner, 1997), pp. 51 and 54.
32 Other responses too are possible as we read such texts. For interesting comments on the part that sadism and masochism might play here, one could look at work on the horror genre, especially Carol J. Clover, *Men, Women and Chain Saws* (London: British Film Institute, 1996; originally 1992). See too Bataille on De Sade, *Eroticism*, pp. 164–96.
33 Creed, 'Horror and the Carnivalesque', p. 156.
34 Vanacker, 'V.I. Warshawski, Kinsey Millhone and Kay Scarpetta', p. 65.
35 Scarpetta is often emotionally affected by the demands of her job, which points again to an ambivalence about the stability of subjectivity and of the social order within the genre. It also points in another and related direction, to a female empathy that cuts against what Vanacker calls 'the male epistemological ideal', 'V.I. Warshawski, Kinsey Millhone and Kay Scarpetta', pp. 80–3.
36 Patricia Cornwell, *Post-Mortem* (New York: Avon, 1990) p. 27. Scarpetta is referring here to random violence in American society but the statement is more generally revealing, given Scarpetta's profession and the materials with which she works so closely.
37 In 'Cultural Work, City Crime, Reading Pleasure', Stewart uses Bataille's arguments concerning the fascination that death holds to explain the readerly attraction to violent excess in crime fiction. Interestingly, Bataille himself goes on to explain the attraction of the detective hero in popular detective fiction in such a context. To 'lay themselves bare to danger' (*Eroticism*, p. 86) is something the strongest men [*sic*] do, following the call of that 'resolute desire' identified with the contemplation of death. Through the detective here, then, 'we live vicariously in a way that our lack of energy forbids us in real life. Without too much personal discomfort we experience the feeling of losing or of being in danger that somebody else's adventures supply' (p. 87). The overdetermination of macabre bodily exploration and dissection and physical decay in Cornwell's texts adds a further problematic dimension to such an analysis.

11
Desires and Devices: On Women Detectives in Fiction

Birgitta Berglund

The traditional pattern of representing women in fiction as objects and men as subjects has in general posed great difficulties for those (presumably female) writers who have wished to create strong and positive women protagonists. Because of the specific demands of the genre, this is even more true of detective fiction. Thus, in spite of the great number of women writers in this genre, it is a fact that the overwhelming majority of detectives in fiction have until quite recently been men. Women in detective stories have been victims, or they have been perpetrators, but they have not, on the whole, been detectives – that is, they have not been given the most important part to play. In novels written by men, women detectives are very few indeed (although they do exist) but even in books written by women, male detectives dominate. Thus we have, for instance, such notable fictional detectives as Hercule Poirot, Lord Peter Wimsey, Albert Campion, Roderick Alleyn, Adam Dalgliesh and Reginald Wexford – all of them created by women.

In *Sisters in Crime: Feminism and the Crime Novel*, Maureen Reddy suggests several reasons for this situation. One is that writers who want to reach large groups of readers tend to choose a male protagonist rather than a female one, as women are on the whole much more willing to read about men than the other way round: 'girls read the Hardy Boys series but few boys would want to be caught with a copy of a Nancy Drew book in their hands,' she comments.[1] Anybody at all familiar with the reading habits of the sexes would have to admit that there is, alas, a great deal of truth in this. Another reason for the lack of women detectives could, according to Reddy, simply be the fact that literature reflects reality, and since most real-life detectives are men it is only natural that the detective story reflects this situation. On the whole, however, this seems a less convincing suggestion, since the detective story in its vintage form is not really a realistic genre. Neither the crimes that are committed nor the criminals that exist in it can be said to reflect reality. As for the detective, neither Sherlock Holmes nor Hercule Poirot or Lord Peter Wimsey can be regarded as realistic or even credible representatives of the profession.

Reddy obviously comes closer to the truth when she notes that 'the whole notion of a woman in charge, and especially a woman presumably dedicated to ideals of law and order, works against traditional expectations.' In other words: the real difficulty in creating a woman detective has more to do with literary patterns and expectations than it has to do with real life. The problem is the fact that the detective in the classic detective story is the typical hero: strong, intelligent, resourceful, a latter-day knight who fights and defeats evil. According to the pattern established by Conan Doyle with Sherlock Holmes, he is also an almost superhuman mastermind who is allowed a great degree of eccentricity and egocentricity because of his extraordinary powers.[2] With such forefathers, what can a woman writer do? Or, to put it more precisely, what could the early women writers of detective fiction do? How could they unite this ideal with a traditional feminine ideal and come up with a credible woman detective? What they had to contend with was the fact that a woman could not be a hero; she had to be a heroine, which is a very different thing. A heroine may be allowed a great deal of love and romance, but she is allowed very little scope for action. Real-life women have worked hard in a great many fields to support themselves and their families for the whole of this century at least, but heroines have not.[3] 'Theirs has been the marriage plot, the erotic plot, the courtship plot, but never, as for men, the quest plot', writes Carolyn Heilbrun, who as Amanda Cross was one of the first to try to change this state of affairs.[4] 'What can a heroine do?' Joanna Russ asked in an essay from 1973, and came to the conclusion that the only really viable options for a heroine were falling in love, going mad or dying.[5]

None of this is of course much use to a detective, and one answer would be that women in books simply cannot be detectives.[6] A woman who lives up to the feminine ideal can really only be suited to the rôle of victim – either to be rescued by the hero in the nick of time or ending up as the pretty, blonde body in the library. A woman who does not live up to the ideal, on the other hand, a woman who is independent, resourceful and assertive, will most probably be cast in the rôle of the perpetrator, the villainess. In fact, if she wears a tight-fitting jumper and occurs in a novel by Dashiell Hammett,

Raymond Chandler or any of their followers in the hard-boiled school, the chances are that she will be both, the villainess *and* the victim, seducing the hero in the first part of the story and ending up as a decorative corpse in the second part. This villainess/victim dichotomy is almost taken for granted in hard-boiled detective fiction. Marty Roth has pointed to one pertinent example: when it is clear that the body found at the beginning of Vera Caspary's novel *Laura* is not that of the main character Laura (an independent, resourceful and assertive career woman), Laura is immediately suspected of having committed the murder. 'It is as if the form determined that if she is not to be the victim, then she must be the villainess.'[7]

Even to writers of more 'soft-boiled' fiction it seems to have been difficult to get away from these stereotypes. When Dorothy Sayers, after having allowed Lord Peter to dominate the action in four of her first five novels, introduced a strong and interesting woman to her readers (in *Strong Poison*, 1930) she did so by combining the two rôles. Standing accused of the murder of her lover, Harriet Vane appears to us in the rôle of the perpetrator; but since she is innocent, she is in fact the victim, the damsel in distress, whom Lord Peter sets out to rescue.

It is worth dwelling a little on *Strong Poison* and Harriet Vane, because although Sayers' choice of the stereotypes is conventional, her use of them is not. Harriet Vane actually has many of the attributes of the villainess. Not only is she a blue-stocking, a self-supporting and successful working woman with a university degree, at a time when very few women possessed academic qualifications, but she has also defied conventions by first cohabiting with her lover and then leaving him when he offered her marriage. The problematic nature of such a heroine is demonstrated by Sayers when she has most of the people in the story (including the judge and the majority of the jury) regard Harriet's character and actions as proof of her guilt, being perfectly convinced that a woman who is capable of living with a man outside marriage is also capable of poisoning him with arsenic – a line of reasoning that would not appear self-evident today.

By fusing and twisting the two female stereotypes, Sayers has to some extent undercut them, and it is to her great credit that she has managed to make Harriet Vane, who is not particularly pretty and certainly not conventionally feminine, a convincing and moving heroine whom readers can respect as well as pity. Still, Sayers has not so far moved away from the two categories of victims and villains, and I would therefore like to return to my question. What could a woman writer of detective fiction do in the first half of this century, if she wanted to write about a woman character who was neither victim nor villain, a person whom both she and her readers could identify with in a more positive way?

Looking at it from this point of view, it is possible to discern how certain devices have been used in order to present interesting women characters without jeopardising the conventional structure of the plot. I will examine the most frequently used of these devices or strategies here, to see how they operate and whether they are still at work.

One very popular strategy, which was used abundantly in the 1930s, is to make the heroine the girlfriend, fiancée or wife of the detective. In this way she can at least have part of the fun – she can help to solve the crime without losing her feminine qualities. With all due respect to Simenon's Mme Maigret (who does not seem to have much fun, though), it seems that the detective's partner, whether portrayed as the plucky young girlfriend or the understanding wife, receives much more attention in stories written by women than in those written by men. This is certainly the opinion of Patricia Moyes, herself a detective story writer with a married couple as her investigators, who furthermore believes that the partner is often a projection of the writer herself. 'I feel sure that when I am at my nicest and brightest, I am extraordinarily like Emmy Tibbett', she writes, adding, 'nobody else can see the similarity, oddly enough, except that we both have dark hair and are fighting to keep our waistlines: but I *know*. In other words, Emmy is the person I wish I was.'[8] The last few words – 'Emmy is the person I wish I was' – are significant. A more or less strong identification combined with an element of wish-fulfilment is in fact apparent in several partner-heroines created by women writers of the Golden Age. An obvious example is Ngaio Marsh's Agatha Troy, who first appeared in *Artists in Crime* in 1938 and who is made to be a painter, as Marsh herself was at that time, only more professional and more successful than Marsh ever was. In this first book Troy is even allowed to finish a particular picture that Marsh herself had once wished to paint.[9]

Agatha Christie's lively and attractive Tuppence, who solves crimes together with her Tommy, may or may not be a projection of Agatha Christie herself, and she occurs in just four of Christie's novels.[10] However, the fact that Christie kept her alive for over 50 years and allowed her to marry, have children and grow old during this time, seems to indicate that she was more real to her creator than either Hercule Poirot or Miss Marple, who go through no such development.

The autobiographical trait is certainly strong in the case of Harriet Vane, who after her debut as villainess/victim in *Strong Poison* advances to the partner rôle in the three books in which she solves crimes together with Lord Peter: *Have His Carcase* (1932), *Gaudy Night* (1935) and *Busman's Honeymoon* (1937). Like her creator, Harriet has an Oxford degree, supports herself by writing detective fiction and even lives in the same flat as Sayers at Mecklenburg Square in London. The similarities go further and deeper than this, however. As Barbara Reynolds puts it, 'Dorothy's own life, her experience as a writer of detective fiction, her personality, her thoughts and feelings, her very way of talking, even her appearance, run parallel with Harriet's, lending solidity and resonance.'[11]

The way in which the partner strategy is used by women writers can be compared to a passage in the autobiography of a woman artist growing up at the turn of the century. In *Period Piece: A Cambridge Childhood*, Gwen Raverat describes how, although she was deeply interested in art from an early age, it never occurred to her that she could become an artist herself. Instead she dreamed about marrying an artist and being allowed to help him: 'Surely, I thought, if I cooked his roast beef beautifully and mended his clothes and minded the children – surely he would, just sometimes, let me draw and engrave a little tailpiece for him.'[12] This is not very different from what many a woman crime writer in the Golden Age did: not daring to fantasise about actually being a detective herself, she fantasised instead about marrying one and being allowed to help him.

Some women were bolder, though, and did put a lady in the leading part. One of them was Baroness Orczy, who published *Lady Molly of Scotland Yard* as early as 1910. Lady Molly was certainly not the only 'lady detective' to occur in print around the turn of the century, nor was she the first, but since she is fairly typical of her kind she may be allowed to exemplify another useful strategy. In this novel, the author seems to achieve the impossible: she has a likeable, loveable, womanly woman who not only solves crimes efficiently and professionally, but actually does so inside Scotland Yard. How did she manage this?

It is of course not possible to know how the Baroness's mind worked, but I have a suspicion that she and her fellow creators of early lady detectives reasoned along the following lines: Question: In what circumstances is it acceptable – even laudable – for a woman to do a man's work? Answer: When the man is not there to do the job, and the woman does it reluctantly and out of a sense of duty rather than for fun, money or fame. In other words, nobody sympathises with the career woman who neglects her household duties because of an interesting job, but everybody sympathises with the poor widow who has to work to support herself and her children – regardless of whether the two do the same work and are equally successful. Likewise, when the country is at war and the men are away, women are expected to run the farms and factories and are appreciated for their efforts to do men's work – always on the understanding that they will gracefully give it all up when the men return. This is the concept and the strategy that Baroness Orczy used. When Lady Molly is asked to take charge of Scotland Yard's 'Female Department', she accepts the position with the sole purpose of being able to solve the murder for which her husband had been unjustly imprisoned and thus prove him innocent. It takes her several years to do it and she manages to solve a great many other crimes meanwhile, but it is the thought that counts: her alibi is perfect.

Another variant of this strategy is to put the heroine in a situation which demands, or at least excuses, that she solves a particular crime. The heroine, or somebody close to her, may be wrongfully suspected or accused of the crime in question, or she may be anxious to clear the name of a deceased relative, or she may have reason to suspect that either she or somebody close to her will be the next victim of the murderer. In all of these situations the heroine will be seen as acting either in self-defence or out of a laudable sense of duty, and she can be allowed to express a feminine reluctance to the task while getting on with it.

This is quite a useful strategy which was used frequently by early writers of detective fiction,[13] but it has one drawback. Because it is built on the heroine's involvement in a particular crime and the assumption that once this is solved everything will return to normality again, this type of heroine cannot easily be made into a series character. As most writers of detective novels, not to mention readers, seem to favour recurring characters who can be allowed to change and develop (like Lord Peter Wimsey) or at least become increasingly well known and loved (like Sherlock Holmes) this is a serious drawback. Lady Molly, once she has proved her husband innocent, disappears from the scene and, in Julian Symons' words, 'presumably settles down again to domesticity'.[14] Symons obviously feels that domesticity becomes the lady better than detection, as he calls Lady Molly 'a woman detective more disastrously silly than most of her kind', a judgement which partly seems based on the time it takes her to solve the original case and expose her husband's wrongful conviction. Symons' attitude here shows that he has not grasped the problem, which is the need to provide the woman detective with an excuse for what she is doing.

Symons points to another problem with fictional women detectives when he complains that the early ones 'retained an impossible gentility of speech and personality while dealing with crime.'[15] This is of course the Catch 22 situation that any woman encounters when trying to make a career in a man's world. If she does not retain her feminine attributes, she is accused of being unwomanly, and if she does, she is accused of being unprofessional. So, to return to the question articulated at the beginning of this essay, what can a woman do? The answer brings me to the most successful device for creating a feminine female detective, one who is not just a female Watson allowed to tag along after the real detective, one who does not need any excuses for her involvement in criminal cases, and who can occur in one book after another without requiring any greater suspension of disbelief than is usual for the average male amateur sleuth: the spinster detective.

The most famous spinster detective is Agatha Christie's Miss Marple, but Christie did not create the type. In fact, she is originally not a creation of the Golden Age at all. The first notable spinster detective is Miss Amelia Butterworth, whom Michele Slung calls 'the prototype of the elderly busybody female sleuth',[16] and who assisted New York police detective Ebenezer Gryce in two books by Anna Katherine Green, *The Affair Next Door* (1897) and *Lost Man's Lane* (1898). But the spinster detective's real rise to fame and popularity came in the period between the wars. Her first appearance then was as Dorothy Sayers' Miss Climpson, introduced in 1927 in *Unnatural Death*. Then came Patricia Wentworth's Miss Silver, a retired governess who first appeared in 1928 in a novel called *Grey Mask*. Finally, Miss Marple made her debut in *Murder at the Vicarage* in 1930, after which there were dozens more on both sides of the Atlantic. 'Dear old tabbies are the only possible kind of female detective', Sayers wrote to Christie on the publication of *Murder at the Vicarage*, and Miss M. is lovely.'[17]

There are several reasons why 'old tabbies' were such a good choice in the 1920s and 1930s, one of them being the historical situation. In the period between the wars there was a surplus of unmarried women in England, so the elderly or middle-aged spinster was a familiar figure whom readers could easily recognize and in many cases no doubt identify with.[18] Ivy Compton-Burnett observed about these women:

> They say that before the first war there were four or five men novelists to one woman, but that in the time between the two wars there were more women. Well, I expect that's because the men were dead, you see, and the women didn't marry so much because there was no

> one for them to marry, and so they had leisure, and, I think in a good many cases they had money, because their brothers were dead, and all that would tend to writing, wouldn't it, being single, and having some money, and having the time – having no men, you see.[19]

All this could also, some writers apparently felt, encourage sleuthing. What is interesting about the spinster detective is that at a first glance she would seem to go against all rules. What we have is a detective who is not only a woman, but also a woman who is neither young nor pretty nor, it would seem, prominent in any other way, but a quite plain, usually badly dressed, ostensibly quite unprofessional, seemingly quite scatter-brained and even slightly ridiculous old maid. We are light-years away from Sherlock Holmes, the mastermind – and that is precisely the reason for her success. The spinster sleuth is so completely harmless and endearing, and so essentially feminine in her ways and manners, that she can get away with murder – or at least the detection of murder – without threatening male authority. She can also get away with several pretty severe proto-feminist statements that would be much more provocative coming from the lips of a more powerful or sexually attractive woman. Within the framework of the story the spinster sleuth can also turn her own low status to her advantage by making people tell her things they would never tell a real detective, because they never suspect her capacity. It could be said that the writers who use this kind of character are playing a double game: they make use of a traditional female stereotype – the ridiculous old maid – in order to subvert it and explode it.

No doubt there was an element of getting one's own back in the creation of the old maid who outsmarts the men around her and not everybody relished the idea. Michele Slung mentions 'a 1941 survey of detective-story readers who listed as their prime bêtes noires "nosy spinsters... women who gum up the plot... super-feminine stories... heroines who wander around attics alone".'[20] However, the strategy was on the whole an extremely successful one, and 'the knitting brigade', as they have sometimes been called, dominated the market for women detectives in fiction well into the 1960s.

During that decade, however, new things started to happen. The women's movement slowly became visible even in the generally conservative genre of detective fiction, and there were writers who scorned the old devices, creating strong and successful women characters who need no excuses or alibis for what they are doing. In other words, they created women who are heroes rather than heroines.

One of the first 'new' women detectives was Amanda Cross's Kate Fansler, who first appeared in 1964 in a novel called *In the Last Analysis*. Kate Fansler is representative of the new breed of women detectives in that she is an independent, successful and self-confident professional woman who detests cooking and cleaning and can afford to do so. This state of affairs is illustrated in a scene which has become almost obligatory for Cross's followers. The protagonist comes home hungry and tired at the end of the day and looks into her fridge – which is, as she well knows, either empty or filled with moulding left-overs. She decides to settle for a whisky or to go out for a meal. Like many of her detecting sisters, Kate Fansler is also an outspoken feminist and some of the mysteries she solves – in particular *Death in a Tenured Position* (1981) – are closely connected with feminist issues. Kate Fansler is also fairly representative in that she is an academic. Several of the new women detectives are academics, as were and are many of the traditional male detectives.[21] Among more recent feminist academic sleuths, Joan Smith's Loretta Lawson might be mentioned. Like Fansler a lecturer in English literature, Lawson made her debut as an amateur detective in 1987 in the aptly named novel *A Masculine Ending*, which among other issues discusses the problems of gender-biased language.

Nonetheless, academics, even if they are staunch feminists, are relatively decorous ladies compared to the really tough women who have recently made their way into fiction. Some of them are policewomen, including Susan Dunlap's Jill Smith who was introduced in *As a Favour* in 1984 and who is an interesting inversion of the 'Lady Molly type' of detective. Jill joins the Berkeley Police Department in order to support her husband through graduate school, after which she intends to settle down and have children. This would seem like an impeccable excuse of the traditional kind, if it were not for the fact that Jill finds she likes the job so much that she stays on and divorces her husband. P.D. James's Kate Miskin of Scotland Yard, who first appeared in *A Taste for Death* in 1987, likewise puts her career before personal relationships. In *Original Sin* we see Kate enjoying the outward sign of her success, a Docklands flat, while showing few regrets at saying goodbye to the lover who wants a stronger commitment than she does. Lynda LaPlante's Jane Tennison, introduced in *Prime Suspect* in 1991 and famous not least because of Helen Mirren's portrayal of her on television, should also be mentioned in this context. If these women seem like embodiments of the career-woman stereotype who would have been the villainess of the Golden Age, this is even more true of

Katherine V. Forrest's Kate Delafield. She is a member of the Los Angeles Police Department introduced in *Amateur City* in 1984, who is not only an extremely strong and tough Vietnam veteran, but also a lesbian who 'comes out' at the end of the second book in the series, *Murder at the Nightwood Bar* (1987).[22]

Finally, the best proof that something new has really happened to the detective story is that women have entered the previously almost exclusively male field of hard-boiled fiction. Probably the most famous, if not the first, private investigator of the hard-boiled school is V.I. Warshawski, the creation of Sara Paretsky, who first appeared in *Indemnity Only* in 1982. Like her previously mentioned colleagues within the police force, Warshawski is notably without excuses or alibis for her choice of profession. She has a degree in law and once worked as a lawyer, but has given up that career as well as marriage to an up-and-coming young man in order to be her own woman. She runs a reasonably successful, if dusty and disorderly, detective agency in downtown Chicago and has a tendency to become involved in cases of murder, fraud and corruption. She is usually badly beaten up at least once in each novel, is keen on baseball, likes to relax at her favourite bar drinking beer or Black Label whisky, and considers it a trial when circumstances force her to put on a pair of tights and a skirt.

Her character is obviously to some extent based on Philip Marlowe and his kind, and she compares her own work to that of Lord Peter Wimsey in several of the books, but Warshawski is also an outspoken feminist who identifies with other women while refusing to be reduced to a female stereotype. In this as in many other respects she resembles Sue Grafton's Kinsey Millhone, who made her debut the same year as Warshawski in *A is for Alibi*, and Marcia Muller's Sharon McCone who preceded her by five years, appearing first in *Edwin of the Iron Shoes* in 1977 – and a number of other tough and successful women who have emerged during the last two decades.

One that particularly deserves mentioning in this respect is Patricia Cornwell's Kay Scarpetta. The first Scarpetta novel, *Postmortem*, came out in 1990, and since then Cornwell has produced roughly one book a year, allowing her protagonist's life and her relations to the secondary characters to change and develop in interesting ways. The fact that Kay Scarpetta is a medical examiner, who not only solves (generally very violent) crimes together with the police but also cuts up the bodies of the victims, might initially seem almost like overdoing the hard-boiled aspect. However, Scarpetta's 'unwomanly' profession proves to be an extremely clever strategic device, since it makes it possible to present her as a professional woman who is regularly involved in murder cases without actually being a police officer.

Is there no longer any need, then, for the old devices to which the writers of the Golden Age had to resort in order to be able to present a woman detective? The answer is no – and yes. On the one hand, the strong and self-confident women academics, police officers and private investigators who choose to solve criminal cases for fun, fame or money obviously need no excuses to do so, and the spinster detective would seem to be gone never to return. On the other hand, the old devices are still in use among writers who present slightly less daring protagonists. Above all, a surprisingly large number of women in detective fiction still act as the detective's partner rather than as the actual detective. I will just mention three examples out of many. One is *In the Shadow of King's* by Nora Kelly, which came out in 1984. The protagonist of this book, Gillian Adams, is an American historian with a position at a Canadian university, who has studied at the University of Cambridge and who has a relationship with a Scotland Yard detective – a circumstance which comes in handy when a person in the audience is murdered while she is giving a lecture. The writer herself is an American historian with a position at a Canadian university who, according to the blurb of the book, 'is herself no stranger to the city and university portrayed in her first novel.' She is obviously also no stranger to her predecessors in the genre and their methods, as the story contains allusions both to the Vane–Wimsey and the Troy–Alleyn relationship.

Janet LaPierre is a former high-school teacher from Arizona who lives in northern California and writes detective stories. Her heroine, Meg Halloran, is a high-school teacher from Arizona now living in northern California, who, in the novel *Children's Games* (1989), develops a relationship with the police officer who is investigating the murder of one of her students. Finally, in Barbara Crossley's *Candyfloss Coast* (1991) the heroine is a journalist in a fictitious seaside town called Northport, who becomes involved in a murder case connected with a nuclear waste scandal and has an affair with a handsome Yorkshire policeman. And yes, Barbara Crossley has worked as a journalist with the *West Lancashire Evening Gazette* in Blackpool since 1976 and won a British Press Awards commendation for her coverage of nuclear discharges into the Irish Sea.

In other words, the partner strategy is still used, and, it would seem, used with the same biographical angle as in the 1930s.[23] In fact, one recent novel, *Oxford Exit* by Veronica Stallwood (1994), manages to combine a rather shaky performance of woman-as-detective with the traditional motifs of woman-as-partner and woman-as-damsel-in-distress. The main character (who, like her creator, is an Oxford-based former library assistant who has recently turned full-time writer) tries to solve a murder case but does not quite make it. She is also attracted to the masculinely handsome police officer who is officially in charge of the case and who, in the nick of time, saves her from becoming the next murder victim.

As for the strategy of providing the heroine with an excuse for doing what is seen as man's work, it has been used by several modern writers, including P.D. James in *An Unsuitable Job for a Woman* as late as 1972. Private investigator Cordelia Gray has an excuse very similar to that of Lady Molly sixty years earlier. An orphan with no money and no formal education, she unexpectedly inherits the detective agency where she has previously been employed, and somewhat reluctantly decides to have a go at it – partly out of a sense of duty and loyalty to the friend who bequeathed it to her. Just like Lady Molly, Cordelia thus neither actively applies for nor even chooses her job. Both are pressed into accepting their jobs by men and both fit into the category of women who have to fend for themselves in the absence of male protectors. Kate Miskin, too, is provided, if not exactly with an excuse, at least with an elaborate explanation for her choice of profession. An intelligent, illegitimate (and orphaned) girl from a poor working-class background simply does not have very many opportunities when it comes to choosing a career – hence her untraditional choice of the police force which, she thinks, will at least give her the opportunity of promotion because of merit rather than connections.

It is also notable that the women who need no excuses are above all to be found in the hard-boiled genre. It seems to be more difficult to fit strong women into the traditional analytical detective story, although there are some examples. I would especially like mention one writer who has managed to create a likeable and credible woman protagonist who is neither a hard-boiled detective or policewoman nor an academic. I am referring to Jill Paton Walsh, whose first detective novel, *The Wyndham Case*, was published in 1993. The story is set in a Cambridge College by the name of St Agatha's (!), it has a body in the library in the first chapter and a protagonist who comes as close to the spinster detective as seems possible in the 1990s. Miss Imogen Quy is a quiet, middle-aged, unmarried woman (her fiancé jilted her after she had given up her medical studies for him), who works part-time as a College nurse and takes in lodgers in the large house she inherited from her parents. She does *not* have an affair with the rather likeable police officer who is called in to solve the case (the possibility is suggested and rejected); but she does indeed solve the case because people tell her things that they would not tell a police officer. Her only 'excuse' for meddling with police work is the fact that she is genuinely affected by what has happened. We see her not only identifying the murderer but also travelling to the north of England to attend the funeral of the murdered undergraduate, trying to console his mother and giving practical help to his pregnant girlfriend. With echoes of both Christie and Sayers in the plot and a discussion between Miss Quy and her lodgers as to which is the better writer, Ruth Rendell or P.D. James, it is obvious that Jill Paton Walsh places her allegiances within the tradition of the classic British detective story. What is interesting about her novel from the point of view of this chapter is the fact that it features a woman detective who is not particularly tough and who may seem to fit in with gender stereotypes rather than challenging them, but who is nevertheless capable and successful.

Even so, the prospects for the detective story and the woman detective in the future are an open question. In *Murder by the Book? Feminism and the Crime Novel*, Sally Munt claims that the feminist crime novel was a temporary thing which had its heyday in the 1980s during the Thatcher and Reagan era and whose popularity is declining in the 1990s.[24] Possibly this is so. However, Sharon McCone, Kinsey Millhone and V.I. Warshawski are still going strong, as are Kate Fansler and Loretta Lawson, and several other women series detectives who first emerged in the 1990s. Imogen Quy first appeared in 1993 and in 1994 P.D. James's *Original Sin* gives rather more space to Kate Miskin than it does to Adam Dalgliesh. It does seem that some of the most interesting and vigorous works in the genre are those written by women about women, and that feminist or not, with or without the old devices, the woman detective is definitely a literary character who has come to stay. If I might venture a prophesy about her future, it is that the border between the soft- and the hard-boiled will become increasingly blurred. The tough private eye will become more vulnerable, perhaps even be allowed to have a family, while the gentle spinster will turn out to have been a feminist all the time; and we will see more 'ordinary' women who juggle families and careers while staying in charge of the case.

Notes

1 Maureen T. Reddy, *Sisters in Crime: Feminism and the Crime Novel* (New York: Continuum, 1988), p. 6.

2 G.K. Chesterton's almost equally influential character, Father Brown, may seem unassuming and human enough, but this is far from the case. In fact, Father Brown is very nearly God. Julian Symons also classes Father Brown among the 'Supermen of detection' in his standard work *Bloody Murder* (London, Sydney and Auckland: Pan, 1992), p. 94. Needless to say, Symons has no superwoman category. T.J. Binyon claims that it is Edgar Allan Poe's Dupin rather than Sherlock Holmes who is the prototype of the great detective, the eccentric genius with stupendous reasoning powers, but the reasoning is the same and the pattern excludes women: *'Murder will out': The Detective in Fiction* (Oxford and New York: Oxford University Press, 1990), p. 5.

3 Nicola Beauman notes the discrepancy between what women did in real life and what they were allowed to do in fiction in her study of women's novels in England between the wars. Even though 'by 1914 women made up one-third of the total labour force, novelists continued to ignore this vast proportion (some five million women) as a subject for fiction. The 146 000 female clerks who declared themselves in the 1911 census remained unlikely material for novels, which were still dominated by the love interest, by the subject of women's lives as, how and when they were arranged by men': *A Very Great Profession: The Woman's Novel 1914–39* (London: Virago, 1983), p. 42.

4 Carolyn G. Heilbrun, 'What was Penelope Unweaving?', in *Hamlet's Mother and Other Women* (London: The Women's Press, 1990), pp. 103–11, here p. 108.

5 Joanna Russ, 'What can a Heroine do? or Why Women can't Write', in Susan Koppelman Cornillon (ed.), *Images of Women in Fiction* (Bowling Green, Ohio: Bowling Green University Press, 1973), pp. 3–20.

6 Julian Symons quotes Ruth Rendell who, as late as the 1960s, clearly felt that it could not be done. She made her detective a man, she says, because she felt that 'one writes about men because men are the people and we are the others': Symons, *Bloody Murder*, p. 223.

7 Marty Roth, *Foul and Fair Play: Reading Genre in Classic Detective Fiction* (Athens, Georgia and London: University of Georgia Press, 1995), p. 120. The misogyny of this genre is well-known. Heilbrun calls it 'the traditional, tough-guy, nonfeminist, anti-woman American detective novel' ('The Detective Novel of Manners', in *Hamlet's Mother*, p. 242), and Roth claims that 'hard-boiled detective fiction is written against women. At the center of the form, the detective discovers, as if in a dream, that the woman he loves kills men: not only is the attractive woman a criminal, but the extent of her depravity equals his investment in her.' (p. 121) Reddy, finally, writes that 'the treatment of women as objects in male hard-boiled detective fiction results in a simple, clear pattern – women are all potentially destructive and predatory, with some women redeemed by their willingness to submit to patriarchal rule' (*Sisters in Crime*, p. 102).

8 Patricia Moyes, 'The Lot of the Policeman's Wife', in Dilys Winn (ed.), *Murderess Ink* (New York: Bell, 1981), pp. 139–41, here p. 139.

9 Ngaio Marsh, 'Portrait of Troy', in Winn (ed.), *Murderess Ink*, p. 142.

10 *The Secret Adversary* (1922), *N or M?* (1941), *By the Pricking of My Thumbs* (1968) and *Postern of Fate* (1973). There is also a collection of short stories featuring the Beresfords, *Partners in Crime* (1929).

11 Barbara Reynolds, *Dorothy L. Sayers: Her Life and Soul* (London, Sydney and Auckland: Hodder & Stoughton, 1993), p. 255.
12 Gwen Raverat, *Period Piece: A Cambridge Childhood* (London and Boston: Faber & Faber, 1988), p. 129.
13 In the Introduction to her anthology of stories featuring women detectives, Michele Slung also observes, with regard to the early ones, that 'it was not uncommon for the woman to have turned sleuth in order to clear the name of a husband or sweetheart, or to seek revenge': Michele Slung (ed.), *Crime on her Mind: Fifteen Stories of Female Sleuths from the Victorian Era to the Forties* (New York: Random House, 1975), p. xx.
14 Symons, *Bloody Murder*, p. 96.
15 Ibid., p. 101. Slung also comments on the early 'lady sleuths that they were usually *over*-endowed with feminine charms to compensate for their mannish profession', *Crime on her Mind*, p. xix.
16 Slung, *Crime on her Mind*, p. xxii.
17 Quoted in Janet Morgan, *Agatha Christie: A Biography* (London: Collins, 1984), p. 196.
18 Amanda Cross, 'Spinster Detectives', in Winn (ed.), *Murderess Ink*, pp. 96–7, here p. 96.
19 Quoted in Beauman, *A Very Great Profession*, p. 6.
20 Slung, *Crime on her Mind*, p. xxv.
21 Binyon mentions 22 men plus Kate Fansler in his chapter on Academics (*Murder Will Out*, pp. 50–6), and Reddy observes that 'judging from the number of crime novels set in academe, murder – committing it or investigating it – would seem to rank just behind research and teaching in the list of professorial activities' (*Sisters in Crime*, p. 42). An interesting variant on this theme, as pointed out to me by Robert Vilain, is the fact than in Reginald Hill's Dalziel and Pascoe novels, it is the policeman's wife who is an academic – and is deeply suspicious of her husband's profession.
22 Kate Delafield is by no means the only lesbian detective in fiction. In fact, lesbian crime fiction is by now a recognised sub-genre, important enough to be given a separate chapter in Sally Munt's overview *Murder by the Book? Feminism and the Crime Novel* (London and New York: Routledge, 1994).
23 Even the hard-boiled girls do it. Kinsey Millhone, Sharon McCone and V.I. Warshawski all have affairs with policemen at some point, although this does not prevent them from carrying on with their own cases; and Crown Prosecutor Helen West in Frances Fyfield's novels has a long-running relationship with a senior police officer. As these strong women of the 1980s and 1990s, who succeed in spite of male prejudice, to some extent seem to represent the same sort of wish-fulfilment as the spinster who outwits the men around her did in the 1930s; it makes one wonder whether there is a large number of women about who dream of having a romantic relationship with a policeman.
24 Munt, *Murder by the Book*, p. 201.

12

A Band of Sisters

Margaret Kinsman

Famous detecting partnerships are a recurring thematic and structural convention of the crime and mystery literary tradition. From Holmes and Watson, through Poirot and Hastings, Harriet and Lord Peter, and on to Easy Rawlins and Mouse, the detective and the sidekick constitute a distinguishing generic feature. Traditionally, the fictional superhuman detective is in need of no-one as friend or assistant; yet paradoxically, while the detective is seen and understood as isolated and apart, he is often part of a pair who sleuth together and have a rudimentary friendship. If one of the generic functions of partners such as Watson and Holmes is to 'help' the detective solve the crime, another function of the sidekick is to help protect the detective's legendary stance of detachment from other people. However, modern crime and mystery fiction writers have modified this generic convention to present increasingly diverse configurations of friendship, exploring that mysterious and universal dynamic that unites people across barriers of age, race, class, distance and experience.

Contemporary writers such as P.D. James, Amanda Cross, Tony Hillerman, Walter Mosley, Colin Dexter and Sue Grafton portray detective protagonists with friends whose personal and professional rôles extend well beyond those exemplified by the sidekick in the works of Conan Doyle and Christie. In contrast to Holmes and Poirot – sleuths whose legendary clear-sightedness is achieved at a distance from people – the likes of detectives Kate Fansler and Easy Rawlins and Chee see things clearly because they are close enough to others to understand what motivates people. It is the purpose of this chapter to suggest that the motif of female friendship in particular, as explored in the private eye novels of Sara Paretsky and Linda Barnes, serves to destabilize some of the traditional generic expectations of detective fiction, placing the oppositional concerns and values of feminism(s) at the centre of the narrative. Thus a central paradox in the genre lends itself to the concerns of feminist writers.

Paretsky's and Barnes's novels comprise a study of contemporary women's issues that emerge out of the continuing relationships portrayed between their private-eye heroines and female friends who function, not to distance the detective, but to knit her into her later twentieth-century urban context. Close readings of the friendship networks of fictional protagonists V.I. Warshawski and Carlotta Carlyle reveal both their creators to be concerned with the frequently competing claims of friendship and professional responsibility, of autonomy and nurturance, of past and present. Both writers explore, through their protagonists and their friends, the possibility of individual and collective female agency and the effects of such agency on patriarchal systems. As Sally Munt suggests, the act of legitimizing female bonds and according literary value to such intimacy is a pioneering development in the overwhelmingly male-characterised mystery fiction genre.[1]

The motif of female friendship is visible in the work of numerous women crime and mystery writers in the USA and the UK. A partial list would include, for instance, Amanda Cross, Margaret Maron, Laurie King, Barbara Wilson, Kathleen V. Forrest, Barbara Neely, Valerie Wilson Wesley, Edna Buchanan, Joan Smith, Michelle Spring, Gillian Slovo, Val McDermid, Liza Cody. The varied treatments of female friendship by such writers serve a number of purposes: they explore themes of difference (cultural, sexual, racial), suggest ways of understanding the relationships between past and present, offer places of resistance to patriarchal constraints, and locate female communities as agencies of change. There is considerable complexity at work in these texts, with much to explore in relation both to genre convention and to the project of reconfiguring female destinies. While the constraints of space (and the inevitable personal preference of the author) limit this chapter's discussion to two white, middle-class and heterosexual protagonists, it is worth noting that an ever increasing diversity of characters' ethnicity, class, age and sexual orientation is making its way into the genre. Although the female friendship motif as explored by Paretsky and Barnes offers key points for commentary in this essay, there are many other contemporary mystery writers whose work would lend itself to similar analysis.

Paretsky's and Barnes's fictional detectives, V.I. Warshawski and Carlotta Carlyle, are grounded in urban communities of women who share strong emotional and professional ties. Connected to highly visible female networks of working women from many walks of life,

V.I. and Carlotta, in their respective cities of Chicago and Boston, make and sustain friendships across divides of experience, age, class, ethnicity, special needs and sexual orientation. Thus the detectives enact what Carolyn Heilbrun has described, in writing about the celebrated friendship between Englishwomen Vera Brittain and Winifred Holtby, as 'the enabling bond that not only support[s] risk and danger but also comprehend[s] the details of a public life and the complexities of the pain found there.'[2] So how does female friendship make a difference to genre convention? A response to this question requires a brief look at both feminism and genre history.

Feminism(s)

Maureen Reddy recently argued that, after a long gap between Dorothy Sayers's *Gaudy Night* (1935) and *In The Last Analysis* (1964) by Amanda Cross (aka Carolyn Heilbrun), the last three decades have seen feminist crime novels, feminist literary criticism and feminism(s) as social movement all 'growing up together' to challenge literary and social convention, and to become identifiable as distinct counter-tradition.[3] While crime and mystery fiction is, as the Afterword to Walker and Frazer's *The Cunning Craft* points out, 'the subject of increasingly intense and varied theoretical inquiry', so is the contemporary explosion of women writers using the genre to feature a variety of strong female protagonists endowed with courage, autonomy, distinction, and feminist values. Feminist literary and cultural studies scholars such as Marty Knepper, Lyn Pykett, Maggie Humm, Kathleen Gregory Klein, Sally Munt and Anne Cranny-Francis have written, since the 1980s, about the new female sleuth and the ways in which she is undermining the myth of male credentials.[4]

Marty Knepper, in an early article on feminism and Agatha Christie, described feminist writing as that which

> shows as a norm and not as freaks, women capable of intelligence, moral responsibility, competence and independent action ... reveals the economic, social, political and psychological problems women face as part of a patriarchal society ... presents women as central characters, as the heroes ... explores female consciousness and female perceptions of the world; creates women who have psychological complexity and transcend the sexist stereotypes.[5]

Kathleen Gregory Klein's 1988 discussion of gender and genre extends Knepper's definition to include writing that 'rejects the glorification of violence ... objectification of sex ... patronisation of the oppressed. It values female bonding.'[6] Carolyn Heilbrun's *Writing A Woman's Life* identifies a specific feminist agenda in the construction of literary protagonists who articulate 'a self-consciousness about women's identity both as inherited cultural fact and as process of social construction' and who 'protest against the available fiction of female becoming.'[7] Heilbrun herself, writing both cultural theory and popular fiction, has made a pivotal contribution to the dual projects of feminist protest and reconfiguration.[8] Lyn Pykett's 1990 discussion of the female sleuth after feminism pays tribute to Heilbrun and other women novelists who 'intervene in particular traditions of detective fiction to explore, exploit and remodel their conventions.'[9]

Like Heilbrun, and in no small measure drawing inspiration from her achievement in the 1960s and 1970s with the feminist academic sleuth Kate Fansler, Linda Barnes and Sara Paretsky and their wise-cracking detectives share a proclivity to outspokenness about patriarchy in its more unseemly manifestations. The authors also share an inclination to write a multiplicity of female voices in their novels, thus locating the quest for truth characteristic of the genre in shared, collaborative experiences. This crucial aspect serves to challenge the authority traditionally accorded in the genre to the singular male detective. In addition, the multiplicity of voices portrayed in Paretsky's and Barnes's series mirrors one of later twentieth-century literature's most salient features – the urge to explore a wider range of stories and to put on record the accounts of those who have historically been confined to the margins of white Western male experience. The device and theme of female experiences of friendship emerge then as transforming agents which owe a good deal to the real experiences of women involved in contemporary feminist movements (political, social, cultural, theoretical) from the 1960s onwards. In an interview about the eighth Warshawski novel, *Tunnel Vision* (1994), Paretsky said of her famous heroine,

> V.I's character is important to me in a personal way. She ages with each book because she started as a person who was molded by specific experiences of my generation. She was shaped by the civil rights movement, the anti-war movement, and the feminist movement.[10]

As the discourses of feminism unarguably have become part of the later twentieth-century political and social landscape, it is no surprise then, that images of women's friendship proliferate in the current surge of crime and mystery fiction, as well as surfacing in other cultural expression such as television's Cagney and Lacey and Hollywood's Thelma and Louise. Feminist scholars' analyses of gender and genre have focused variously on the development of the female private eye, on the sub-genre of lesbian crime fiction, on race, on social issues as they affect women and on treatments of violence and sex. Paretsky, along with Cross, Muller and Grafton, is generally considered one of the founding pioneers of the feminist private eye, paving the way into the hard-boiled tradition with the 1982 debut of her famous sleuth, V.I. Warshawski. Linda Barnes came on board with ex-cop and licensed private investigator Carlotta Carlyle in 1987, joining a significant group of women consciously writing feminocentric plots and characters with pro-women sensibilities. Klein's discussions of Paretsky and others in *The Woman Detective* and her preface to *Twentieth Century Crime and Mystery Writers* call attention to the considerable part their work has played in shifting crime and mystery fiction from its definitional pedestal of masculinist values. Munt's discussion of feminism and the crime novel in *Murder by the Book?*, in common with Klein and Reddy,[11] considers why the form proves so attractive a vehicle for oppositional politics. She sees Paretsky's liberal feminist stance as limiting, suggesting that while the surface text (the new style protagonist and women-centered plots) may look progressive, in effect, 'radical content is derailed by the deep structures of conventionality undermining it.'[12] It can be argued, however, that V.I.'s own place in the text, often in conflict over her duties to family members and friends, or reflective about her own urges towards violence, or morally ambivalent about how to proceed in her investigations points to a deeper engagement with conventional structures than Munt allows. Munt's recognition that 'traditional form is broken by having [this] female friendship at the narrative centre' offers a starting point for the subsequent discussion.[13]

Susan Koppelman points out in the introduction to her 1991 anthology of stories about women's friendships, 'women have always cherished friendships with women, and women writers have always told stories in which women's friendships are central realities.'[14] She comments further on the complexity and significance of literary female friendships, considering both the features that unify (ethics of equality, reciprocity and mutual support), and those that divide (racism, class oppression, jealousy, betrayal, cultural notions of propriety, status and competition). The formal and thematic trope of female friendship in the crime and mystery fiction genre is as yet a relatively under-explored area of scholarly enquiry, although contemporary studies of friendship in literature can be found in Nina Auerbach's *Communities of Women* and Janet Todd's *Women's Friendship in Literature*.[15] In a genre more noted for portraying women as victims, *femmes fatales* or 'intuitive' sidekicks, female friendship is emerging as a significant theme and structuring device, offering a counterpoint to the genre norms of active men and passive women, and injecting a vitality into the canon of crime and mystery fiction. The subversive and transgressive nature of such counterpoint becomes clearer in the context of some historical aspects of the mystery genre.

Genre history

The origins of the sleuth and the assistant device can be traced back to Poe's 1840s 'Chevalier Dupin' stories. In the relationship between the intellectually superior Dupin and his anonymous friend who narrates the stories can be seen the blueprint of an abiding structural convention of the formula: the superhuman sleuth assisted by a trusted, but less able sidekick, who does the legwork, keeps order (domestic and otherwise) and tells the story from a perspective of admiration and utter loyalty. The Dupin stories also offer the model of an enduring thematic treatment of friendship, the claim of friendship as a rationale for the detective's initial involvement in the investigation. In 'The Purloined Letter', for example, it is Dupin's 'old acquaintance, Monsieur G____, the Prefect of the Parisian police' who prevails upon Dupin in the name of friendship to solve the problem of the missing letter.

By the 1890s, Arthur Conan Doyle was successfully incorporating similar narrative devices in his popular stories of the supremely intelligent Sherlock Holmes and the awestruck, if uncomprehending, auxiliary Dr Watson. The pattern has prevailed ever since. Most fictional detectives have their Watsons in one form or another, and a succession of acquaintances and relations who request the great detective's help, in the name of friendship or kinship.

This traditional sleuth/sidekick configuration has served more than one purpose in the genre. The prototypically eccentric, peculiar, intellectually superior and egotistical detective has long been associated, by readers and commentators, with a deep inner loneliness. By definition something of an isolate, fictional investigators, from Victorian Paris and London to America's mean city streets, have not been easy characters to befriend or live with. Paradoxically, the mythically aloof and self-reliant detective figure has rarely been portrayed without the support of a helper of some description. For the conventionally male, celibate and reclusive detective, this friend often anchors the sleuth into a semblance of domestic routine and familiarity. The faithful assistant and friend figure then is both an expression of the detective's problematic stance *vis-à-vis* society and representative of the social order that he is restoring.

For many decades, the sleuth/sidekick relationship, ostensibly based on friendship, relied less on principles of mutual benevolence, reciprocity and emotional involvement, than on a hierarchical model of human interaction based on acknowledged superiors and inferiors. This inequality further ensured that the powerful bonds of friendship rarely developed to the point where they might compromise the detective's legendary heroism and power. The sort of personal accountability implicit in a close relationship was clearly an impediment to the generic detective's habit of placing himself in potential jeopardy in order to further the investigation. The detective historically remained immune to the ordinary feelings, passions and weaknesses experienced by those around him. The sleuth's foil was characteristically a person who, out of his own lack of knowledge and his excess of emotion, neither stole the great detective's thunder nor drew attention to any possible frailty or stupidity in the super sleuth.

The emerging private eye tradition of the American writers of the 1920s and 1930s featured the urban investigator working alone, critical of society, distrustful of friends, taking comfort in cynicism and bourbon. Here the Watson figure metamorphosed in different directions. One was in the private investigator's uneasy relationship with the agencies of law, in the form of police and judiciary. For instance, Chandler's Marlowe sustains a strained friendship with Bernie Ohls, his ex-partner and now Chief Investigator with the Los Angeles DA's office. The relationship veers between trust and acts of professional assistance, to suspicion and rivalry. Another was in a gesture toward a somewhat more egalitarian comradeship as featured in husband and wife detecting teams; the urbane, wise-cracking, gin-swilling Nick and Nora Charles, for example.

The 1930s appearance of the 16-year-old Nancy Drew and her chums, Bess and George, offers a contrast to the hard-boiled developments, and as Heilbrun asserts, 'a moment in the history of feminism.'[16] Heilbrun cites the early Nancy Drew works as 'the model for early second-wave feminists' who began writing mystery fiction in the 1960s and 1970s (Cross, Muller, Uhnak and Paretsky, for example). Nancy Drew's kinship and support system conforms to certain genre conventions and challenges others. She is motherless; the admiring Bess and George do not always figure it out as quickly as Nancy does; Nancy prefers to sleuth alone, knowing that friends' and family's protective instincts only slow her down. However, she also hints at what was to come in Heilbrun's second wave. Nancy has unusually mature and mutually respectful relationships across age, employment and parental boundaries with both housekeeper Hannah Gruen and her father, incorporating a rare sense of negotiated and chosen obligation and filial accountability. Nancy's autonomy and mobility, augmented by her beloved roadster and by her genuine feelings of affection for Bess, George and Ned which nonetheless do not constrain her, are noteworthy historical features in the genre.

Meanwhile, the Golden Age writers of Great Britain (Christie, Allingham, Marsh and Sayers) were moving their traditionally remote detective figures into less anti-social and eccentric configurations. All four authors develop their male protagonists into more introspective characters who become connected to society in both kinship and friendship systems. As the American hard-boiled detective becomes more characterized by cynicism and sleazy hotel apartments, the Golden Age hero becomes less estranged from the ordinary networks and commitments that weave people into their society. Ngaio Marsh's police detective Roderick Alleyn's intellectual superiority and upper-class social isolation are mitigated by his marriage to the equally intelligent and professionally accomplished Agatha Troy. The eventual addition of a young son Ricky allows Alleyn to demonstrate shared parental concerns. Dorothy Sayers used most of *Gaudy Night* to relate, in addition to a mystery, the crucial struggles of Harriet Vane and Lord Peter Wimsey to achieve a mutual understanding of an egalitarian marital partnership. Agatha Christie's cozy amateur investigating duo Tommy and Tuppence Beresford offers another literary exploration of a marriage partnership based on principles of friendship.

As a structuring device then, traditional sidekick figures such as Watson or Poirot's long-suffering Hastings provided the writer with a means of illustrating both the possibilities and the limitations of the scientific, methodical reasoned approach embodied in the detective and of heightening suspense. In common with the fictional Watson, the reader 'sees' thing differently from the detective. By contrast with the sleuth's superior reasoning powers, the dazzled friend's diligent questions and guesses, slower and more mundane but at the same time illustrative of intuition and imagination, include the reader. Further, the detective/sidekick scheme mirrors the symbiotic relationship between the detective and the criminal. The structural device of such pairings also helps the writer to establish the appearance of hierarchical order and the semblance of social continuity.

All of this is disrupted when female friendship, in the shape of detectives with feminist values and a critical mass of similarly-inclined friends, enters the frame. Carlotta's and V.I.'s places in the text as bold women with initiative, vitality and a taste for adventure already signal disorder and confusion in a genre given to privileging male agency and concerning itself with the stereotypical female destiny of victim or villain. Discussing feminist detective fiction, Maggie Humm notes, 'traditional detective novels distrust friendships and emphasize the security of stable class and geographic boundaries ... a group of novels by feminist women is now challenging the gender norms of detective writing.'[17] The mystery writer's rôle, shared with the fictional detective, is to explore what goes on beyond the façade or behind the framework of any society. The mystery is famously two stories – what happens and what appears to happen. Novels by Paretsky and Barnes also place at the narrative centre stories of what happens and what appears to happen to strong women characters struggling to find a place and a voice within patriarchal social systems. 'Ours is a society which glorifies the masculine and fears the feminine', said Paretsky in a 1990 public lecture. She went on to say, 'if we cannot speak, or if our efforts at speech are not understood [the] omission denies women a chance to feel mastery, to feel competence, to feel confident that they can succeed ... every aspect of history is studied and told with a strong bias against the stories of women, indeed against the value of women.'[18] The difficulties of locating female agency and voice in a society, and a genre, where agency is associated with maleness cannot be underestimated.

It is clear that newer slants on both the theme and device of friendship are taking writers and their detectives into different territory, where not only do the detective and their friends notice the racism, sexism and 'class-ism' that infuses Western society, they obliquely criticise it, even as they solve the crime. By the 1990s the genre abounds in fictional sleuths engaged in complex and challenging relationships. Innovative pairs such as Walter Mosley's Easy Rawlins and Mouse, or Hillerman's Navajo policemen Chee and Leaphorn, or Kellerman's Peter Decker and Marge Dunn blur previously clear boundaries between the rôles signified by the detective and the sidekick. These partnerships tell stories of cultural, social, ethical, religious difference between friends, confronting the antagonisms and the benefits that come with friendship. Where the detective and the sidekick once signified and reinforced inequality in the social status and methodological approach of two individuals, cultural diversity and difference are now likely to feature in friendship networks centred on the sleuth. Historically, fictional partnerships tended to mirror in one way or another the boundaries constructed by society in relation to class, race, gender, sexuality and age. Contemporary writers are more reflective of social and cultural trends to loosen such boundaries. The work of women writers with a desire to make women feel powerful has been key to this development in the genre.

Female friendships

As the genre continues to diversify, the novels of writers such as Paretsky and Barnes reflect some of the more innovative feminist impulses in contemporary society and culture. If the kinds of people we love, respect, admire and accept responsibility for tell us as much about ourselves as individuals and as a community, then there is much to explore in the friendship networks of V.I. and Carlotta. The new wave of female protagonists first demonstrated that women could do the job of detective, and, like Nancy Drew, could have adventures. The next step this critical mass of protagonists has taken has been to re-affirm the value of affiliation and connection for the detective, who is increasingly linked to families of origin, families of selection and friends. Showing a greater sensitivity to the needs of families, friends and children – and to the crimes that threaten them – has not compromised the contemporary female sleuth's toughness or gumption. In each of Sara Paretsky's novels, the irascible, but deeply loyal, Vic winds up reluctantly seeking help from friends and neighbours. Her self-mocking quest to sleuth alone and sleuth fast is valiantly undertaken and is often portrayed as a losing battle in the face of the implacable, ever-present Lotty and Mr Contreras. Almost every Paretsky plot includes a concern for one or more of V.I.'s extended family members, however conflicted and problematic V.I. may find this.

Female friendships as they are written by Barnes and Paretsky illustrate women electing and sustaining peer relationships, even when the friendships are constructed across barriers of age or experience. The age-differentiated friendships between Lotty and V.I., for example, and between Carlotta and Paolina, do not rely on a subordination model, but owe more to social worker Jane Addams's early twentieth-century concept of the construction of the civic family, as she lived it in her famous Chicago settlement, Hull House. These detectives are located in a predominantly female network of elective and affiliative relationships that taken together construct a family of choice rather than a family of origin. The result is a detective whose individuality is established in an identity distinct from familial definition and who also collaborates in the construction of a sex-group identity which asserts its difference from male experiences of the world. The cultural notion of what constitutes a 'proper family' is increasingly interrogated, marking a change from filiative to affiliative authority and telling stories of negotiated intimacy, care and responsibility.

V.I., Carlotta and their assorted women friends are consistently portrayed as actively engaged in the public sphere, working as lawyers, public servants, businesswomen, doctors, artists, political aspirants and lobbyists. They are economically independent, though frequently without much margin. Often they are politically involved in causes related to social justice for women and children. They are possessed of physical and mental strength and stamina, which they work hard to sustain. These detectives have friends and family that take up their time and attention, drawing the reader's attention to the perennial mystery of how to 'be' an adult, achieving, professional woman in a patriarchal world that regards female agency and strength with consternation at best and hostility at worst.

While Virginia Woolf reminds us that we think back, if we are women, through our mothers, Carolyn Heilbrun reminds us of the literary power of the motherless heroine, for whom other destinies are possible because the mother, whose mission it is to 'prepare the daughter to take her place in the patriarchal succession' is absent.[19] Paretsky and Barnes place their orphaned protagonists in transgenerational friendships with a mentoring aspect, so that female destinies other than those dictated by misogynistic strictures are validated, authenticated and enabled. Both offer examples of the generational space that often allows friendships of ease and freedom. In these friendships, moral codes are hammered out and tested. Both V.I. with Lotty and Carlotta with Paolina have to work out an ethics of friendship that takes into account the needs of self and of others, and that can be flexible and responsive to changing circumstances. This challenges the deeply embedded genre construct of the singular, abstract and judgmental code of the avenging knight (detective), answerable to no-one and set apart from his society.

Instead, through the device of female friendship, the genre investigates one of our culture's abiding social constructs for females: the 'ethic of responsibility' as Carol Gilligan's *In A Different Voice* (1982) describes it.[20] Gilligan's famous psychoanalytical argument identifies the affiliative and co-operative connections with others which women are taught to value, and which marks the difference in female ways of thinking and decision making. This ethic, in the words of Maureen Reddy, 'places a concern for relationships [familial and otherwise] ahead of such abstractions as order and justice', and perceives relationships 'in terms of balancing needs'.[21] Awkward characters who query the notion of female selflessness, as Vic does when she is confronted in *Burn Marks* with her indigent and alcoholic Aunt Elena, also throw into relief the ethics of a society that treats the poorer and ageing segments of its population as throw-away. In an anguished dialogue with herself, we see Vic's dilemma mirroring cultural assumptions about sacrificing, nurturing women: 'I was tormented wondering how much I owed my aunt. Would my uncle Peter thrash in guilt for saying no? Of course not.... Did I have a duty to Elena that overrode all considerations of myself, my work, my own longing for wholeness?'[22] Paretsky's exploration of the conflict between society's expectation of the ministering angel and a woman's resentment and guilt is effectively conveyed in V.I.'s conviction that if she acts from duty, she will feel a want of freedom.

V.I.'s closest, most significant friendship is with the self-possessed Dr Lotty Herschel, an older woman of great personal and professional stature, dignity and compassion. Lotty, a Holocaust survivor, has practised medicine in Chicago all of her adult life. She and V.I. meet as pro-choice activists in the early 1970s, organising clandestine safe abortion referral networks at the university. They share an analysis of oppression; a commitment to action; and a vision of a different, more just society. Lotty's and V.I.'s friendship is portrayed as a complex and changing one, characterized by both intimacy and space. They get over many a personal crisis together because they are able to acknowledge mistakes and to incorporate change. Lotty, a mentor rather than a Watson, is the friend who aids the mysterious search for self more than the search for the solution to the crime. In turn, V.I. becomes a revered and put-upon rôle model, much to her dismay, to the younger Caroline in *Toxic Shock*. By the end of the novel, V.I. has come to an uneasy truce with Caroline over the demands and expectations of their relationship, which dates from their childhood together in a working-class South Chicago neighbourhood.

The limits, boundaries and possibilities of both new and old friendships are constantly tested in V.I.'s world. While Paretsky leaves little room for sentimentality, her treatment of V.I.'s relationships with Lotty, the redoubtable Sal ('Sal's a shrewd businesswoman. The [Golden] Glow is only one of her investments')[23] and other characters offers scope for investigating an ethics of friendship. In *Burn Marks* for example, V.I. struggles to reconcile a growing awareness of her old friend and comrade Ros Fuentes's corrupting political ambition with Vic's instinct for female solidarity and her political commitment to seeing more women take up public office. Vic also seeks to reconcile her desire for personal and professional autonomy with the strong sense of personal loyalty which implicates her in gestures of reciprocity and obligation. This results, for example, in V.I.'s often-expressed frustration at Lotty's expectations and demands for accountability, even while V.I. appreciates Lotty's unwavering emotional and intellectual companionship, and practical medical support. The extent to which Vic is able to extend to others the autonomy she desires for herself is put to a severe test in *Guardian Angel*, when Lotty's safety is jeopardised by an unthinking act of V.I.'s. Lotty withdraws from Vic in anger and pain, and it costs Vic dearly to allow Lotty the space and time to heal herself. The relationship between the two of them seems to have cracked beyond repair under the strain of V.I.'s recklessness and Lotty's subsequent intransigence in refusing to countenance such lack of consideration for others. 'The longer Lotty and I went without speaking, the harder it was going to be to get back together.'[24] It takes considerable courage and determination from both of them to find a way to reconcile their differences, forgive their trangressions and reaffirm their deep connection to each other. It is a suspenseful and convincing rendition of the ups and downs of committed friendship.

Barnes casts the mentoring rôle of female friendship across differences of age and culture somewhat differently, though again without sentiment and with an ethical framework. Private eye Carlotta's volunteer work as a Big Sister brings her a friendship with the younger Hispanic Paolina, growing up uncertain and poor in the mean streets of a Boston slum. This relationship is explored and developed through the series of novels, with a constant awareness of the social power structures that place Carlotta in an advantageous postion in relation to both Paolina and her struggling, suspicious mother. While the potential abuse of power and privilege is never forgotten by the writer, Carlotta and Paolina's bond clearly suggests that such relationships also foster good things for both participants. As Carlotta puts it 'We've been spending Saturday afternoons together forever. We cruise the shopping malls, check out the local music scene, go apple picking in the country. I've taken her to eight Red Sox games, and she really got into it last year.'[25] The older woman detective and the young schoolgirl learn from their not unproblematic friendship that growth and change are possible, intimacy is legitimate, dependency is a virtue, self-reflection is enabling and fear can be smart. Reflecting on her commitment to Paolina, who is curious about the identity of her drug-baron absent father, Carlotta says, 'lies don't usually bother me much, but I try not to lie to Paolina. She means too much to me. And lies have a sneaky way of tiptoeing back to haunt you.'[26]

In both series, the protagonists' array of friendships, which embrace pluralistic constructs of sexual, cultural and ethnic identities, are portrayed as relationships rooted in a basic respect for the other. With no agenda of subordination, the strengths, weaknesses, fallibilities, uncertainties and rages of both self and other can be explored and accepted as if in a safe harbour.

As well as foregrounding the idea of women mentoring each other and working to establish an ethics of affiliation, the device and theme of female friendship shifts the loner detective figure into new patterns of kinship and community. Domestic arrangements of the detective and her friends, for example, may illustrate experiments in living other than the traditional nuclear family or the eccentric spinster model so often encountered in literary works. Carlotta lives in a Boston house inherited from an aunt, who is one of a number of strong maternal forebears in Carlotta's life. In a gesture towards both economy and community, Carlotta shares the house with the quirky punk-ish Ros, whose erratic housekeeping skills relieve Carlotta of that particular burden, and whose energy and humour often invigorate and inspire Carlotta. V.I., who claims sloth and a short fuse as two of her outstanding characteristics, shares her Chicago apartment with no-one on a permanent basis. Lotty also lives alone and relishes her privacy. As V.I. herself points out, 'some men can only admire independent women at a distance.'[27] Both Vic and Lotty, who value domestic solitude as an active choice, are able to live alone without taking on the unremitting isolation normally associated with the single female. No recluse, V.I. shares her dog Peppy with the downstairs septuagenarian neighbour Mr Contreras, which involves her in constant, tricky negotiations as to what constitutes appropriate neighbourly behaviour. Mr Contreras admits more than once that he prefers being a friend and neighbour to the volatile V.I. than to taking up his jealous daughter's insistent offer of a place to live with her dull suburban family. In these examples of living arrangements, we see illustrated a society in which traditional forms of living are crumbling, alongside visions of what might be appropriate and satisfying replacements.

Carlotta's relationship with her sometime employer Gloria, who owns a cab company, and V.I.'s friendship with Sal, who reads the *Wall Street Journal*, are examples of female friendship anchored in the public domain. These associations resist the legacy of female friendship located in the private, largely unexamined and relatively powerless domestic sphere. A legion of friends from Vic's law school and public defender days occupy offices and boardrooms the length and breadth of Chicago. Carlotta's housemate Ros has her finger on the pulse of the contemporary performing and visual arts scene in Boston. They conduct the business of their lives in a multiplicity of settings, which encompass the kitchen and the living room as well as the law offices, the restaurants, the community service organisations and other bastions of power which constitute the public sphere.

As the sleuth and her friends acquire background and history and biography, the genre begins to suggest new ways of interpreting the connections between women's lives past and present. V.I. and Carlotta both pay homage to powerful matrilineage. V.I. dreams insistently of her mother Gabriella, whose aspirations to become an opera singer were abandoned when she fled fascist Italy to arrive as a young immigrant in Chicago and married the Polish cop Tony Warshawski. Carlotta often invokes the Yiddish sayings of her grandmother, acknowledging Yiddish as 'the voice of exile'.[28] Carlotta attributes her glorious red hair to a great grandmother. She also has fond memories of Aunt Bea, 'an awe-inspiring woman',[29] who paid off her mortgage shortly before she died, leaving the house to Carlotta. The stories of such ordinary women are usually absent from the literary record. Including their stories, as remembered by V.I. and Carlotta, elaborates female kinship and friendship from different generations and different perspectives, with appreciation of the differences as well as the similarities among the voices.

Foregrounding female friendship in feminist detective fiction serves many purposes in the larger enterprise of refiguring female destinies. As the motif of female friendship circulates in the genre, and is itself revisited, revised, extended, interrogated, it begins to reshape the crime and mystery stories themselves. The common narrative device of the partner, while indeed present in both Barnes and Paretsky, locates female friendship as a source not only of those functions we have come to expect of the sidekick (information, protection, refuge, sounding board), but increasingly as a place of revision of received plots and assumptions about women's lives. The singular 'other-half' of Holmes's Watson and Dupin's narrator shifts to a network of affiliation in a feminist sleuth's life, so that the sidekick rôle is transformed into a wider and more transgressive motif. What radical messages are understood and received is an open question, because so much depends, not only on friendship, but on the reader. Where literary courtship usually ends in marriage and the detective's quest ends with the crime resolved, literary friendship's goals are harder to pin down. For V.I. and Lotty, and Carlotta and Paolina, the plot of friendship is a continuing one of co-operative endeavour, pointing towards a future that includes and values many more women's stories and voices.

Notes

1 Sally Munt, *Murder by the Book? Feminism and the Crime Novel* (London: Routledge, 1994), p. 11.
2 Carolyn Heilbrun, *Writing a Woman's Life* (New York: Ballantine, 1988), p. 100.
3 Maureen Reddy, 'The Feminist Counter-Tradition in Crime: Cross, Grafton, Paretsky and Wilson', in *The Cunning Craft: Original Essays on Detective Fiction and Contemporary Literary Theory*, ed. by Ronald G. Walker and June M. Frazer (Macomb, Illinois: Western Illinois University Press, 1990), p. 174.
4 Ibid., pp. 188; Marty S. Knepper, 'Agatha Christie, Feminist', *The Armchair Detective*, 16.4 (Winter 1983), pp. 389–406; Lyn Pykett, 'Investigating Women: The Female Sleuth after Feminism', in Ian Bell and Graham Daldry (eds), *Watching the Detectives* (London: Macmillan, 1990), pp. 48–67; Maggie Humm, *Border Traffic: Strategies of Contemporary Women Writers* (Manchester and New York: Manchester University Press, 1991); Kathleen Gregory Klein, *The Woman Detective: Gender and Genre* (Champaign, Urbana: University of Illinois Press, 1988) and the Preface in *Twentieth Century Crime and Mystery Writers*, 4th edn, (Chicago: St James Press, 1996), pp. vii–ix; Munt, *Murder by the Book?*; Anne Cranny-Francis, *Feminist Fiction* (Cambridge: Polity Press, 1990).
5 Knepper, 'Agatha Christie, Feminist', p. 389.
6 Klein, *The Woman Detective*, p. 201.
7 Heilbrun, *Writing a Woman's Life*, p. 18.
8 See also Carolyn Heilbrun, *Hamlet's Mother and other Women* (New York: Columbia University Press, 1990).
9 Pykett, 'Investigating Women', p. 49.
10 John Ambrosia, 'Sara Paretsky expands her "Vision"', *Oak Leaves* 8 June 1994 (Chicago: Pioneer Press) B4–5.
11 Maureen Reddy, *Sisters in Crime: Feminism and the Crime Novel* (New York: Continuum, 1988).
12 Munt, *Murder by the Book?*, p. 58.
13 Ibid., p. 11.
14 Susan Koppelman (ed.), *Women's Friendships: A Collection of Short Stories* (Norman: University of Oklahoma Press, 1991).

15 Nina Auerbach, *Communities of Women: An Idea in Fiction* (Cambridge, Mass. and London: Harvard University Press, 1978); Janet M. Todd, *Women's Friendship in Literature* (New York: Columbia University Press, 1980).
16 In Carolyn Stewart Dyer and Nancy Tillman Romalov (eds), *Rediscovering Nancy Drew* (Iowa City: University of Iowa Press, 1995), p. 11.
17 Humm, *Border Traffic*, p. 185.
18 Sara Paretsky, 'Women, Speech and Silence', unpublished speech delivered at the University of Chicago, 1990, pp. 9 and 16.
19 Heilbrun, *Writing a Woman's Life*, p. 119.
20 Carol Gilligan, *In a Different Voice: Psychological Theory and Women's Development* (Cambridge, Mass. and London: Harvard University Press, 1982).
21 Reddy, 'The Feminist Counter-Tradition', pp. 183 and 177.
22 Sara Paretsky, *Burn Marks* (New York: Delacorte, 1990), p. 197.
23 Ibid., p. 83.
24 Sara Paretsky, *Guardian Angel* (New York: Delacorte, 1992), pp. 291–2.
25 Linda Barnes, *Coyote* (New York: Delacorte, 1990), p. 35.
26 Linda Barnes, *Cold Case* (New York: Delacorte, 1997), p. 12.
27 Paretsky, *Guardian Angel*, p. 30.
28 Barnes, *Cold Case*, p. 7.
29 Barnes, *Coyote*, p. 34.

13

An Urban Myth: *Fantômas* and the Surrealists

Robert Vilain

Fantômas is the hero, anti-hero, or just possibly virtual hero of 32 *romans feuilletons* by Pierre Souvestre and Marcel Allain written between February 1911 and September 1913.[1] After Souvestre's death from Spanish influenza in February 1914, Allain wrote further stories, in serial and cartoon form, which appeared between 1926 and 1963. According to the terms of the original contract, signed on 29 April 1910 with the publisher Arthème Fayard, Souvestre was to write a book a month for a fee of 2000 francs per volume, with an additional royalty of three centimes for each copy over 50 000 and three per cent of the income from any new editions.[2] Allain, ten years younger, had met Souvestre when he joined the staff of Souvestre's automotive magazine *Le Poids-Lourd* in 1907; they both also worked for the sporting daily *L'Auto* and the entertainment magazine *Comœdia*. Allain was subcontracted to Souvestre for 500 francs per manuscript and half the royalties. Soon they were selling 600 000 copies a month at 65 centimes a copy. The two would sit down together for three days and work out the plot and structure; then they drew lots to see who was to write each chapter. Both dictated their sections into a dictaphone. The chapters beginning *néanmoins* were by Souvestre, those with *toutefois* in the first sentence were by Allain – although the adverbs were usually cut out at the typing stage. It usually took a week to dictate and type up and then two days to proofread.[3]

Books written at such speed and in such an apparently mechanical manner might not initially look very promising material for serious study. Those arguing 'for' in the debate about whether the detective story can claim literary credentials might be tempted to brush the *Fantômas* novels under the carpet, or worse. Fred Bourguignon condemned the novels as 'bon à brûler'.[4] But *Fantômas* became an icon for the literary avant garde in France in the 1910s and 1920s. Robert Desnos several times referred to the series an 'épopée moderne', maintaining that *Fantômas* 'surgit comme un des monuments les plus formidables de la poésie spontanée'.[5] Guillaume Apollinaire described *Fantômas* as, 'au point de vue imaginatif, une des œuvres les plus riches qui existent'.[6] Blaise Cendrars called the series 'l'Enéide des Temps Modernes' and Jean Cocteau praised 'le lyrisme absurde et magnifique de Fantômas'.[7] In 1912 Apollinaire and Max Jacob even formed a kind of fan-club, 'Société des Amis de Fantômas' with Maurice Raynal and Picasso. Artists, too, were inspired. René Magritte used a still from the *Fantômas* films for *L'Assassin menacé* (*The Assassin Threatened*, 1926–7); *Le Retour de flamme* (*The Backfire*, 1943) is virtually a copy of the front cover of the first *Fantômas* volume, with a rose in the man's hand rather than a dagger; *Le Barbare* (*The Barbarian*, 1928) shows the head of Fantômas emerging through a brick wall. The Spanish cubist Juan Gris painted *Fantômas* (*Pipe and Newspaper*) in 1915, where a *Fantômas* novel is shown on a table among familiar, banal objects, including a newspaper well known for its sensationalist coverage of crime.

The reasons for the enthusiasm of such distinguished practitioners of 'high culture' are perhaps not obvious to modern readers – even Marcel Allain was later to write, 'pour moi, *Fantômas* était une bonne affaire mais qui m'a beaucoup amusé.... Je suis étonné qu'on y attache tant d'importance',[8] and the authors deliberately lampooned the avant garde in volume 22 of the series, *Les Amours d'un prince* (*The Affairs of a Prince*), with a spoof journal *Littéraria*. Since *Fantômas* also seems relatively little-known to today's Anglo-Saxon readers, therefore, this essay both offers a flavour of the novels and their main characters and gives an account of some of the responses to Fantômas among the Surrealists and their contemporaries, concluding with some hypotheses about why he should have been so attractive to them.

It has been suggested that the popularity of Fantômas derived partly from the degree to which the novels were a 'tableau naturel et social de la France à la veille de 1914'.[9] Desnos wrote of the series that

> toute la période qui précède la guerre y est décrite avec la précision qui concourt précisément au phénomène lyrique. Intrigues internationales, vie des petites gens, aspect des capitales et particulièrement de Paris, mœurs mondaines telles qu'elles sont... mœurs policières et présence, pour la première fois, du merveilleux propre au XXe siècle, utilisation naturelle des machines et des récentes inventions, mystère des choses, des hommes et du destin.[10]

In its range of geographical settings and social milieux the series is comparable to an extent with Balzac's ambitions in *La Comédie humaine*: it includes the provinces and Paris (as well as the wider world abroad),[11] high society, industry, the church, a circus and the down-and-outs of Montmartre, and very much more. Allain was later to remark,

> un des trucs de Fantômas que je reprends dans tous mes romans ... consiste à toujours situer l'action du roman dans un décor réel. Parce qu'ainsi, le lecteur a toujours l'impression que 'nom d'un chien, s'il était arrivé à ce moment-là!'[12]

Verisimilitude as the basis for fantasy is a tried and tested device and the illustration on the front cover of the first volume and on the posters that covered France prior to publication exploits this, too. It is of a masked man in top hat and tails, leaning earnestly and mysteriously forward, astride the whole of Paris. He is holding a glinting dagger in his right hand, an image for which Robert Desnos claimed 'une importance énorme dans la mythologie et l'onirologie parisiennes'.[13] In fact, the illustration was lifted from a rag-bag of rejected drawings for other publicity campaigns and was originally designed to advertise *Pillules pink pour personnes pâles – Pink Pills for Pale Persons*. Here in place of the dagger there had been a little pill-box out of which were strewn the pink pills, again across the whole of Paris. In the case of the medicine the message was that the pills were a panacea; advertising *Fantômas*, it meant that his influence would be felt in every nook and cranny of the great capital – which was a prospect intended to thrill.

Fantômas, variously called 'l'Insaisissble, le Maître de l'Effroi, le Roi de l'Epouvante, le Tortionnaire' and 'l'Empereur du Crime',[14] is characterized above all by cruelty and violence so heartless as to be almost sublime. He is *so* wantonly destructive, in fact, that Louis Chavance was provoked to write an essay on the morality of the *Fantômas* books in sheer stupefaction.[15] In the eighth novel, *La Fille de Fantômas* (*Fantômas's Daughter*), the 'hero' poisons a few rats during a cruise on the liner *British Queen* and watches while most of the 500 passengers die of the plague. In *Le Fiacre de nuit* (*The Night Cab*) Fantômas sadistically ruins the department store Paris-Galeries because its owner has had the audacity to pay court to his daughter: perfume is replaced by acid in the atomizers, the grand staircase collapses on the Saturday of the sales and the unfortunate Chapelard is finally strangled with an iron gauntlet from his own collection of medieval armour. In *La Fin de*

Fantômas (*The End of Fantômas*) he traps an entire circus audience in a safety net and proceeds to rob them before flinging them bodily aside, and in *Juve contre Fantômas* (*Juve Against Fantômas*) he not only sends a boa constrictor up to the detective Juve's room to squash him to death but derails the Orient Express, killing most of the passengers. More gruesome still, in order to prove that the insurance broker Hervé Martel is not Fantômas in *La Mort de Juve* (*The Death of Juve*), he has Martel executed by an armless [*sic*] avatar of Jack the Ripper who performs the ugly deed with a specially prepared top hat. Ribonard, who tries to cheat his boss of the proceeds of a robbery in *Le Magistrat cambrioleur* (*The Burglar Judge*) is substituted for the clapper of a church bell and dashed to death as the faithful are called to mass. Fantômas is no respecter of youth and beauty either, for in *L'Évadée de Saint-Lazare* (*The Escapee from Saint-Lazare*) after the head of Blanche Perrier is crushed in a mangle, Fantômas slips her scalp into Juve's pocket during a tram-ride.

The crimes are therefore usually as sensational as they are brutal, and this accounts for some of the enormous popular appeal of the narratives. The misdeeds are painted with a broad brush, so breathtakingly nasty that they verge on the anarchic, and there is evidence to suggest that the population of France in the second decade of the century was particularly hungry for whiffs of radical destabilisation. The gang of murderers and terrorists known as the 'Bande à Bonnot' began their series of awful crimes in December 1911 and they were an obligatory feature of the newspapers for months on end, with long columns speculating on the strategies devoted to their capture. Souvestre and Allain were themselves inspired, if that is the word, by the activities of the Bonnot gang and had Fantômas emulate them in *L'Assassin de Lady Beltham* (*Lady Beltham's Murderer*) in July 1912. If it was excessive to denounce *Fantômas* before Parliament in the same year as partly responsible for the wave of terrorism, his exploits were certainly in tune with contemporary interest in anarchism and terrorist crime.

At the other end of the scale, there is also an element of snobbish escapism about the tales. The illustration depicts Fantômas in evening dress, and although his Protean being recombines its molecules to take the form of bartenders, roadsweepers, tramps and accordionists, he belongs, if he can ever really be said to exist, in the upper echelons of society. His thefts are invariably of the jewels worn by society ladies, his dupes include members of the royal houses of Europe such as Frederick-Christian II von Hesse-Weimar, and he is probably the father of Prince Vladimir, aka Comte d'Oberkampf, aka Vicomte de Pleurmatin.

His mistress is the chic English noblewoman Lady Beltham (prophetically homophonous with 'belt 'em'), an *agent provocateur* and accomplice to many of the atrocities in the salons and ballrooms of Paris. She has Fantômas murder her husband, and he manages to substitute another of her suitors for himself when he is captured and guillotined. In this respect Fantômas is part of the gentleman burglar tradition, a bloodier version of the Pink Panther or Raffles, combined with a bit of Scarlet-Pimpernel-gone-to-the-dogs. 'They seek him here, they seek him there, those Frenchies seek him everywhere; is he in heaven, is he in hell? That demmed elusive Pimpernel!' With the right rhyme, Baroness Orczy's celebrated verse could easily be adapted to Fantômas.

Given the swashbuckling exploits of Sir Percy Blakeney, the Orczy comparison is apt enough, but a parallel in the most famous work by the Comte de Lautréamont, *Les Chants de Maldoror*, takes us directly into the area of the Surrealists' connections with popular crime fiction. Maldoror is capable of the wildest metamorphoses and Lautréamont – in reality Isidor Ducasse – describes him thus:

> Aujourd'hui il est à Madrid; demain il sera à Saint-Pétersbourg; hier il se trouvait à Pékin. Mais affirmer exactement l'endroit de ce poétique Rocambole, est un travail au-dessus des forces possibles de mon épaisse ratiocination. Ce bandit est, peut-être, à sept lieues de ce pays; peut-être, il est à quelques pas de vous.[16]

Rocambole was another bandit hero, the extravagant star of Ponson du Terrail's nineteenth-century series of *romans feuilletons*. Rocambole may lie behind elements of the character of Fantômas, but it is unlikely that Allain and Souvestre knew Maldoror well, if at all. For the Surrealists, however, Lautréamont was an icon. Breton gives him an honourable mention in his 'Manifeste du surréalisme',[17] for example, and wrote the introduction to an edition of *Les Chants de Maldoror* (*The Cantos of Maldoror*) in 1938. Magritte contributed an illustration and went on to complete a further 77 for another edition in 1948.[18] Lautréamont's collages and collocations of unpredictable or hitherto incompatible images were inspirational for the Surrealists; the extreme lyricism of his poetic prose, and the way in which he attempted to marry the terrible with the sublime are re-enacted to an extent in *Fantômas*, and it may be that the Surrealists saw the latter as a twentieth-century reincarnation of Maldoror. 'Magnétisant les florissantes capitales, avec une fluide pernicieux, il les amène dans un état léthargique où elles sont incapables de se surveiller comme il le faudrait. État d'autant plus dangereux qu'il n'est pas soupçonné': written in 1869, but true of the 1911 character, too.[19]

There are many other strands of literary tradition into which Fantômas may be woven: some have likened him to Sebastian Melmoth, others to Balzac's Vautrin. There is a Balzac echo picked up by one of Fantômas's many admirers, Pierre Véry, who has Antoine La Guépie, the hero of his novel *Danse à l'ombre* (*A Dance in the Shadows*) meet up with Fantômas for a while. Fantômas soon abandons La Guépie when he catches sight of his arch enemy the policeman Juve in the crowd: 'Et maintenant dit-il "A nous deux, Juve"' – which is an echo of the end of *Le Père Goriot* when Eugène de Rastignac looks out across Paris and says ominously, 'A nous deux maintenant'.[20] What Fantômas has that is not shared to anything like the same extent by these rivals and partners, with the exception of Maldoror, is his extraordinary metamorphic versatility. This is not the same as Sherlock Holmes's capacity for perfect disguise – so perfect that he often fools 'even' Dr Watson – because Holmes always reappears as the same pipe-smoking, fiddle-playing bachelor with an interest in different types of cigar ash. Fantômas, on the other hand, is consumed by his disguises. At the simplest level he suppresses the hapless innocents who cross his path and assumes their identities – he is particularly fond of impersonating rich jewellers and bankers (or, on one occasion, the ghost of a banker), but in *L'Arrestation de Fantômas* (*The Arrest of Fantômas*) he is first the Captain of the *Skobeleff*, then Fred Bolton, the red-head who resides in the Britannic Hôtel *and* at the same time the brown-haired Archinet operating from the Hôtel de la Paix, then a hunchback and an anonymous hairdresser, and finally he emerges as the Tsar of Russia. At the end of *L'Assassin de Lady Beltham* he gives himself up, but can afford to do so because he is simultaneously somewhere else and therefore at liberty, an impossibility that is never explained. Sometimes he puts himself in a position to free himself, adopting the identity of the judge in *Le Magistrat cambrioleur* to ensure he is acquitted of the crimes he is committing elsewhere as Fantômas, giving the order himself for the release from prison of 'Fantômas' (who is Juve in disguise). As the American Tom Bob, in charge of the Council of Five in Scotland Yard, he runs the investigation into the case of Dr Garrick (a pseudonym for Tom Bob, who is still also Fantômas) who is accused of murdering his wife, and proclaims his own innocence. Despite strong suggestions that Fantômas is 'really' Gurn, former artillery sergeant from South Africa, there never emerges a coherent identity from behind the disguises.

Appearing as often as not in a black hooded cape, he is not real in any recognisable social, psychological or even traditionally fictional sense. Every time he is captured, imprisoned, shot at, burnt up or otherwise physically impaired, it turns out that he is not really there, absent *à la* Macavity, that he had escaped in the nick of time, or the authors often reveal that a substitution has been made at a point much earlier in the story. If one returns to check, re-reading the text trying to find the sleight of word that lets him slip away, one cannot find it, because it is not there. The inconsistencies are simply ignored, and Fantômas vanishes almost literally between the lines, remaining a source of terror because his is an identity perpetually deferred and displaced.

There are four more main characters in all the novels. First, there is Lady Maud Beltham, sometimes operating as the Grand-Duchess Alexandra of Hesse-Weimar, or Mme Gauthier, crooked treasurer of a charitable society, or the louche night-club owner Mme Rosemonde. Although she is horribly killed and/or elaborately buried on a number of occasions, her repeated resurrection suggests that the authors were reluctant to sacrifice a strong female character. Next is the detective, Juve of the Sûreté, ostensibly devoted to the Cartesian, ratiocinative approach to detection, but a man so unhealthily obsessed with the capture of Fantômas that he fakes his own death at one point to pursue him undercover. Like Fantômas, he is a devotee of disguise, although nearly always in the Holmes manner, with a simple change of clothes and some good make-up to pass him off as an old crone, a jobbing electrician or the sinister Job Askings, English cutpurse. On occasion he disguises himself as Vagualame, overlapping therefore with one of Fantômas's own favourite masks. On one occasion he is arrested by his own boss on suspicion of being the arch criminal himself, but Juve's insights in fact often successfully uncover Fantômas. He arrests him more than once, but is usually worsted, even to the point of being killed, again more than once. Juve is a passenger on the ill-fated *Gigantic* (sunk, like the *Titanic*, by an iceberg), and on the last page of Volume 32, *La Fin de Fantômas*, Fantômas sensationally reveals that the master detective is in truth his very own twin brother.

The third of the stock characters, the *Capitale* journalist Jérôme Fandor, functions as a kind of side-kick for Juve, not quite a Watson or a Captain Hastings, because he can be relied on to take plenty of initiative himself, on occasion as intrepid as Tintin. He suffers violent setbacks alongside Juve (whereupon he will exclaim 'Crédibicêtre'), and carries off similar triumphs, saving Hélène when she is injured by a falling safe or about to become the victim of an experimental blood-transfusion.

Hélène Gurn herself, sometimes considered the daughter of Fantômas, sometimes not, is a true wild child *avant la lettre*. She was brought up in Natal as a boy, Teddy, becomes romantically involved with Fandor, whom she nevertheless tries to murder occasionally, and is therefore the source of some tension, torn between him and her father. (However, if Etienne Rambert is Fantômas, as the first novel suggests, then Fandor, whose real name is Charles Rambert, is Fantômas's son and his relationship with Hélène is incestuous.) Hélène, too, often disguises herself to her advantage, but more often she is a pawn in his nefarious projects or the victim of his enemies, kidnapped, mistreated, even stabbed. Nonetheless, at the end of *Le Cerceuil vide* (*The Empty Coffin*) she is revealed as none other than Archduchess Hélène de Mayenbourg, niece of the Queen of the Netherlands and heir to her throne. In Chapter XV of *La Fin de Fantômas*, however, she is pronounced granddaughter and heir to the throne and fabulous wealth of Bedjapour, hundred-year-old monarch of all the Indies.

None of these characters is memorable as a personality; there is very little psychological characterisation – and one certainly cannot say of the *Fantômas* novels what John Bayley has said of Sherlock Holmes, namely that they can be read for the sheer joy of the characterisation, never mind the intrigue. The authors are not shy about their contrivances, and while there are gestures made towards realism (the minutely detailed evocations of everyday activities in the city such as tying a shoe-lace or buying a drink), there is little in the way of psychological verisimilitude on a larger scale. These are not like the cartoon characters who are squashed by a bus or a tumbling rock and then pop back into shape like rubber dolls – Tom and Jerry are pushed through many mangles but there is never so much blood as when Blanche Perrier suffers the same fate – but the main protagonists are all mysteriously indestructible and Allain and Souvestre rarely bother to fill in the awkward details of how they manage to reappear.

The plots of these novels are unmemorable. They operate not as closely reasoned chains of action, cause and effect worked out in detail, careful hints dropped by the authors at strategic points to enable the 'crime' to be solved. Instead they are episodic, lurching from predicament to predicament, each scenario ultimately designed to show how elusive and clever Fantômas is and how ultimately futile are the efforts of Juve to catch him. There is plenty of suspense, but it is suspense as to *how* rather than uncertainty about *what*. We know full well that Fantômas is lurking behind anything out of the ordinary that happens, and we are on the edge of our seats waiting to see which innocuous bit-part turns out to be the arch-villain himself and what atrocity he will perpetrate to get himself out of a tight corner. Overarching this play of peek-a-boo there is a general schema that is common to most, if not all the novels. Max Jacob described it thus, tongue-in-cheek:

> *Première partie*: description d'un milieu paisible ou non. Les auteurs font preuve d'une connaissance minutieuse du milieu décrit. Cette description est large et simple.
>
> *Deuxième partie*: un crime est commis avec des circonstances mystérieuses généralement par un domestique faux et dans un endroit hermétiquement clos. L'habileté du policier Juve découvre que l'auteur du crime est un complice de Fantômas.
>
> *Troisième partie*: l'habileté du policier Juve découvre qu'une haute personnalité (quelquefois un monarche) n'est autre que Fantômas grimé.
>
> *Quatrième partie*: une course fantastique amène Fantômas poursuivi et le policier poursuivant sur divers points du globe: chute du Niagara, tuyau de locomotive, environs de Grenoble, rade de Cherbourg, ferrures extérieures de la Tour Eiffel, le fond de l'Atlantique, etc. Dans ces occurences, les auteurs ne se montrent pas seulement des savants très forts en géographie et en 'Guide Joanne', des chauffeurs experts, mais des poètes aigus, terribles, des auteurs dramatiques d'une ingéniosité unique.[21]

Jacob adds another key comment, an observation made by every subsequent critic, namely that 'le style est déplorable, tranchons le mot, il est déplorable'. The authors commit crimes against grammar and syntax, and, most heinous of all, 'ils abusent du mot "désormais".'[22] In other words, Jacob is acknowledging that the series that he has just lauded in another part of his homage, via a poem devoted to the exploits of his hero, is at root the purest pulp fiction.

Jacob's poem is headed 'Ecrit pour la S.A.F.', the *Société des Amis de Fantômas*. Jacob was later to write two exquisitely ironic prose poems on Fantômas for his Cubist collection *Le Cornet à dés* of 1945. Apollinaire wrote several times about *Fantômas* in the *Mercure de France*, lauding the series as 'plein de vie et d'imagination, écrit n'importe comment, mais avec beaucoup de pittoresque'.[23] Partly because of the films made of *Fantômas*, it attracted 'un public *cultivé* qui se passionne pour les aventures du policier Juve, du journaliste Fandor, de Lady Beltham, etc.'.[24] Between April 1913 and July 1914 Louis Feuillade, who worked for Gaumont, made five of the novels into hour-long films. They starred René Navarre, who as a consequence was mobbed in the streets whenever he showed his face. The films are fast-moving, immensely detailed in their reconstructions of location, often with a kind of urban lyricism as Fantômas leaps from rooftop to rooftop making his escape. But Apollinaire's 'public cultivé' were won over before the films appeared, and these merely reinforced the appeal of the books themselves.[25] When Apollinaire adds of the books that 'ce sont même, je crois, à peu près les seuls livres que j'aie bien lus et j'ai eu le plaisir de rencontrer nombre de bons esprits qui partageaient ce goût avec moi',[26] he is making a sophisticated point about the particular intellectual and cultural appeal the series had for writers of his ilk. He praises the books for their richness, their precise description and for the mine of information they will contain for future scholars of early twentieth-century *argot*. While this is true, it is also tongue-in-cheek and Apollinaire is perfectly well aware that the real merit of the novels has little to do with future PhDs in Linguistics.

The hyperbolic enthusiasm was not confined to the S.A.F. Blaise Cendrars included a poem on Fantômas in his *Dix-neuf poèmes élastiques*. Robert Desnos, when he was young, was fascinated by the covers of the novels (which one critic has called the only really poetic thing about them), and his interest in the figure of the chameleon–villain has been linked to his study of the alchemist Nicolas Flamel. Both he and Cendrars were certainly striving to make something magical out of the objects of their world, something to replace the lost magic and myth of past cultures, and they may have seen in Fantômas a kind of magus. On the other hand, both were susceptible to a good joke, too, and Cendrars later worked on a film script combining Fantômas, Arsène Lupin and his creator Maurice Leblanc, Conan Doyle, Gaston Leroux's Rouletabille in a kind of Keystone Cops detective-chase movie. Desnos is the author of perhaps the most famous homage to Fantômas, the 'Complainte de Fantômas', twenty-six stanzas of six lines each, rhymed ABBACC, romping through a series of his most famous exploits in abbreviated form. It begins thus:

Ecoutez … Faites silence …
La triste énumération
De tous les forfaits sans nom,
Des tortures, des violences
Toujours impunis hélas!
Du criminel Fantômas.

> Lady Beltham, sa maîtresse,
> Le vit tuer son mari
> Car il les avait surpris
> Au milieu de leurs caresses.
> Il coula le paquebot
> Lancaster au fond des flots.[27]

And so on, self-consciously hushed and devoted, yet making much play of contrived rhymes, shortened or elided syllables, and some fairly unsubtle rhythmic devices to enhance the ironic pathos. The 'Complainte' was broadcast on Radio Paris at peak time, set to music by Kurt Weill, with Antonin Artaud as artistic director. Attracted by the peculiar blend of lyricism and drama, Desnos was also to try to make an opera libretto out of *Fantômas*. Although his plans were never fully realized, they reveal the underlying seriousness of his preoccupation with the novels: he focused on 'l'équilibre entre deux forces identiques et contraires' in Juve and Fantômas, making central the revelation that the two are twins and extending it to the elemental, almost existential energies that the novels encapsulate. Fandor and Hélène replay the conflict on a different level; Lady Beltham represents 'l'énigmatique et tragique allégorie de l'amour, de ses passions et de ses désordres' and the various villainous minions who support Fantômas in his career of crime are 'l'humanité en proie à la faim, aux vices, aux maladies et qui triomphe enfin le jour où Juve et Fantômas sont morts. Ils détruisent l'ordre ancien.'[28]

There are many other tributes from the Surrealists and their successors, none as witty as the *Complainte*, I think, and none of really striking quality, but the question remains as to why he should have been so fascinating for them. It cannot have been traditional literary values, since in terms of plot, character and style, Allain and Souvestre have never earned any credit for what one might loosely call literary merit. Indeed, more often than not they have been dismissed as writers of trashy pot-boilers for the public. Some critics have maintained that Fantômas would not be remembered at all if it were not for the historical fact of the Surrealists' attention just before the war. Even so, there are several reasons why the Surrealists were not receptive to just any old crime novels, but to Fantômas in particular, and what follows is an attempt at a summary.

First, there is the dream-like quality of much of the narrative; the feeling that, for all the realistic detail of situation and sometimes of gesture, one's feet are never firmly planted on the ground, that the action is propelled almost gratuitously and that its discontinuities verge on the insane. Breton's manifesto defines Surrealism as 'la croyance à la réalité supérieure de certaines formes d'associations … à la toute-puissance du rêve, au jeu désintéressé de la pensée'.[29] The Surrealists' use of dream, trance and non-rational states, where the world was ordered, if at all, then analogically rather than logically, will have found a counterpart in the Allain/Souvestre narrative precisely because their mode of composition was so sloppily sensationalist, operating less by carefully motivated cause-and-event sequences than by the overriding principle of moving on and keeping up.

Second, the siting of the novels, for the most part, in modern Paris.[30] This is no longer the gloomy, sinister Paris of Eugène Sue, but a cosmopolitan, rapidly moving, anonymising metropolis in which banal familiar landmarks, streets, hotels, buildings and so on, receive a mysterious and slightly numinous patina from their contact with the hero. The city is well on the way to gaining its own modernist identity in these novels.

Third, the hero's non-identity that I described earlier. Fantômas is almost a spoof on the gradual decomposition of individual identity that characterizes one strand of French fiction in the later nineteenth century. Naturalism and Impressionism dominated in the 1880s and 1890s, and the two movements are fundamentally united in that they pay attention to the phenomena that reach us via our senses and not to the core, essence, centre – call it what you will – that earlier generations took for granted. Rimbaud's half-despairing, half-triumphant cry of 'Je est un autre' is the consequence of this, enacting and promoting a search elsewhere for the co-ordinates of the self that was picked up by the Surrealists, most directly by Desnos. Magritte, too, focuses on this aspect in Fantômas:

> He is never entirely invisible. One can see his portrait through his face. … His movements are those of an automaton; he brushes aside any furniture or walls that are in his way. … Fantômas's science is more valuable than his language. We do not guess and cannot doubt his powers.

Juve's attempt to capture his adversary is described thus:

> Fantômas is close by, sleeping deeply. In a matter of seconds Juve has tied up the sleeper. Fantômas continues to dream – of his disguises, perhaps, as usual. Juve, in the highest of spirits, pronounces some regrettable words. They cause the prisoner to start. He wakes up, and once awake, Fantômas is no longer Juve's captive.

> Juve has failed again this time. One means remains for him to achieve his end: Juve will have to get into one of Fantômas's dreams – he will try to take part as one of its characters.[31]

Fantômas, by not existing within even the familiar co-ordinates of Paris, of a body with a stable exterior, within the moral co-ordinates of crime and punishment, is another step in the flux, the indeterminacy and the radical destabilization of the self that was one of the Surrealists' starting points.

Fourth, and perhaps more properly a sub-category of one of the other reasons, there is the whiff of anarchy that attaches to Fantômas. He fascinates and attracts by virtue of his sheer evil, and if he can toy with the fate of the whole of the Paris aristocracy, escape society's restraining measures, its values and its standards then he performs a function akin to *épater le bourgeois*, and – a piquancy that will have appealed to the Surrealists – does so despite the fact that the bourgeois were buying the books at the rate of more than half a million a month. Fantômas is a corrupting force, and gleefully so.

Fifth is the glee, a self-perpetuating pleasure in the novels being just as they are, a love of the game. They are a romp and the less high-minded Surrealists will have rocked with pleasure as they read. Roger Shattuck, in *The Banquet Years*,[32] shows how the Surrealists had had the rug pulled out from beneath their world and had therefore to dance to keep their balance, and it is easy to see how Fantômas might have appealed as a model here.

The sixth point is the way in which Fantômas shares in and then subverts the genre of crime writing. A number of precursors and traditions with which one feels compelled to associate 'le Maître de l'Epouvante' have already been mentioned. The villain as hero is a tradition that goes back to Bluebeard and beyond, and one could complete the literary placement by looking forward to the kind of book Fantômas anticipates – Patricia Highsmith's Ripley comes to mind at once. Furthermore, were there not the structure of criminal-plus-accomplices versus policeman-plus-sidekick, the novels would fit more properly into the genre of adventure stories, with perhaps Dumas Père's *Le Comte de Monte-Cristo* as an antecedent. After all, Edmond Dantès takes his revenge on the banker Baron Danglars almost as cruelly as Fantômas treated his victims, but in the adventure story, the hero is always ultimately on the side of the angels, which Fantômas most decidedly is not. The novels invite comparison with a structure that sets out to re-establish order after a bit of fun – when the policemen are *supposed* to catch the criminal – and simultaneously refuses to join in with the tradition: they are obviously much more 'Pink Panther' than 'James Bond'.

Seventh is a familiar but unpredictable tendency for individuals, communities, and particularly intellectual communities, to make an icon out of something diametrically opposed to themselves. Madonna has become one of the icons of the gay community; university undergraduate common rooms elect the stars of television soap operas to honorary memberships; Proust's Swann and the real live Goethe were both involved with unlikely and, in the eyes of the world, unsuitable women, and in Swann's case, one 'qui n'était pas mon genre'.[33] What is at issue here is not just the familiar feeling of 'taking a break' from the high seriousness of one's professional life or aesthetic preferences. There is an element of camp pleasure in being overtly *déclassé* from time to time and the Surrealists probably felt something of this when they latched onto Fantômas.

Finally, when Jean Cocteau wrote of the 'lyrisme absurde et magnifique de Fantômas', he located the source of his interest in the very triviality of the genre, a triviality that paradoxically reveals primordial truths:

> De longue date, j'ai cherché refuge contre la littérature poétique ou réaliste dans les livres où l'auteur ignore que poésie et vérité le hantent, le soulèvent au-dessus de lui-même et dirais-je, du mépris qu'il affiche pour un genre qu'il imagine indigne de sa plume, alors qu'il se trompe en visant beaucoup plus haut. Nous en eûmes, Apollinaire et moi, la preuve, lorsque les auteurs de *Fantômas*, étonnés par notre enthousiasme, nous firent savoir qu'ils étaient capables d'écrire des ouvrages moins naïfs. Or, ces ouvrages moins naïfs étaient d'une naïveté déconcertante. Ils nous démontraient une fois de plus la primauté d'un génie pareil à celui de l'enfance, génie que les personnes mal instruites du collège des muses confondent avec la maladresse et les chances du hasard.[34]

Desnos also surmised that Fantômas was 'un idéal romantique pour beaucoup de garçons' and 'un idéal romanesque et sentimental pour beaucoup de petites filles et même adolescentes'.[35] For Cocteau, however, the appeal goes even deeper, for beneath the unresolved moral play of good and evil there is a childlike innocence to the writing that he finds paradoxically profound.

Raymond Queneau wrote in 1950, 'je voulais écrire une *Vie de Fantômas*', but confessed his inadequacy for the task: 'j'ai lu quatre fois les trente-deux volumes; la cinquième fois, je me suis arrêté au Tome 24'.[36] As a preparatory study for his *Life*, he says, he has collected some statistics for each of the thirty-two novels, counting up the incarnations of the hero, the escapes, the murders, the failed murders, the thefts and so on. He notes that Fantômas botches the murder in almost 25 per cent of the cases; and furthermore that his success rate improves dramatically in the last sixteen volumes, up to 87.5 per cent. It is a spoof, of course. How he would have laughed at the article in *Europe* analysing the type, frequency and context of the verbs for killing in the *Fantômas* series.[37]

Notes

1 The original 32 volumes were reissued with some alterations by Allain between 1932 and 1934; stories 1–22 appeared, abridged, as Pierre Souvestre and Marcel Allain, *Fantômas*, 11 vols (Paris: Robert Laffont, 1961–66); stories 21–32 were reissued, uncut, as Pierre Souvestre and Marcel Allain, *Fantômas*, 3 vols, ed. Francis Lacassin (Paris: Robert Laffont, 1987–89). There is a full bibliography giving all the titles and details of their various re-editions at the end of volume 3, although the volume numbers of some of the 1961–66 series are incorrect. At the time of writing (1999) excellent bibliographical details, an iconography and some critical comment are available on an Internet site devoted to the series (*www.fantomas-lives.com*).

2 The contract is reproduced in the special *Fantômas* volume of *Europe*, 590/591 (June–July 1978), 49–50 and in Jean-Luc Angot, *Fantômas revient*... (Le Côteau: Horvath, 1989), pp. 16–17. Angot's volume contains a substantial collection of *Fantômas*-inspired literature; I am additionally indebted to his account of the genesis of the *Fantômas* series.

3 Both adverbs mean 'nevertheless' and the effect of the alteration must have been something like the spoof radio panel game 'Good News/Bad News'.

4 'Fit only for burning.' Fred Bourguignon, 'Bon à brûler', in *La Tour du Feu*, 87/88 (December 1965), 8–10.

5 'Modern epic... towers as one of the most formidable monuments of spontaneous poetry.' Robert Desnos, 'Imagerie moderne', in *Nouvelles Hébrides et autres textes 1922–1930*, ed. Marie-Claire Dumas (Paris: Gallimard, 1978), p. 458. See also Desnos's remarks in *Les Rayons et les Ombres. Cinéma*, ed. Marie-Claire Dumas (Paris: Gallimard, 1992), p. 84.

6 'From the standpoint of the imagination, one of the richest works in existence.' Guillaume Apollinaire, *Œuvres en prose complètes*, ed. Pierre Caizergues and Michel Décaudin (Paris: Gallimard, 1993), III, 215.

7 'The *Aeneid* of the modern day': Blaise Cendrars, in *Soirées de Paris* (13 June 1914). 'The magnificent and absurd lyricism of Fantômas': Jean Cocteau, 'Préface' to Gaston Leroux, *Le mystère de la chambre jaune* (Paris: Le Livre de Poche, 1960). Quoted in Angot, *Fantômas revient*, pp. 53 and 55.

8 'For me, *Fantômas* was a good piece of business that amused me very much at the same time... I am amazed that people attach so much importance to it.' Quoted in Angot, *Fantômas revient*, p. 69.

9 'A tableau of the nature of French society on the eve of the 1914 war.' Georges Sadoul, *Le cinéma français, 1890–1962* (Paris: Flammarion, 1962), p. 412. See also Claude Dauphiné, 'La Société dans "Le Policier apache"', *Europe*, 590/591, 112–19.

10 'The whole period that precedes the war is described there with exactly the kind of precision that aspires to the phenomenon of lyricism. International intrigues, the lives of ordinary people, views of great cities and Paris in particular, society life as it really is ... the ways of the police, and for the first time the presence of the fantastic in a fitting form for the twentieth century, the unforced use of machines and recent inventions, the mystery of things, of mankind and of destiny.' Desnos, 'Imagerie moderne', p. 458.

11 See Maurice Dubourg, 'Fantômas voyage', *Europe*, 590/591, 79–95.

12 'One of the little tricks of the *Fantômas* books that I use in all my novels ... consists in always setting the action against a real background. This way the reader always has the impression, "Blimey! if only he'd arrived at that moment!" The next sentence rather debunks the serious artistic aim suggested so far: 'Et puis cela impressionne les femmes' – 'And besides, it impresses women.' Quoted in Angot, *Fantômas revient*, p. 30.

13 'Enormous importance for Parisian mythology and oneirology.' Robert Desnos, 'Imagerie moderne', pp. 458–9.

14 'The Elusive One, the Master of Fear, the King of Terror, the Torturer and the Emperor of Crime' – *Fantômas*, passim.

15 Louis Chavance, 'La Morale de *Fantômas*', *Europe*, 590/591, 64–7.

16 'Today he is in Madrid; tomorrow he will be in St Petersburg; yesterday he was in Peking. But to say exactly where this poetical Rocambole is to be found is a task well beyond the strength of my clumsy ratiocination. This bandit is perhaps seven leagues away from this land; perhaps he is a few steps away from you.' Lautréamont/Germain Nouveau, *Œuvres complètes*, ed. Pierre-Olivier Walzer (Paris: Gallimard, 1970), p. 222.

17 André Breton, *Œuvres complètes*, ed. Marguerite Bonnet (Paris: Gallimard, 1988), p. 328.

18 See Suzi Gablik, *Magritte* (London: Thames & Hudson, 1970), pp. 48–61.

19 'Magnetizing vibrant capital cities with a pernicious liquid, he brings them into a lethargic state where they are incapable of taking care of themselves as they should. This is all the more dangerous a situation because it is unsuspected.' Lautréamont, *Œuvres complètes*, p. 222.

20 Véry: 'And now he said, "It's war between us, Juve"'; Balzac: 'It's war between us now'.

21 '*First part*: description of an environment, peaceful or otherwise. The authors prove their detailed knowledge of the place described. This description is broad and simple. *Second part*: a crime is committed in mysterious circumstances, generally by a fake servant and in a hermetically sealed place. The skill of the policeman Juve reveals that the perpetrator of the crime is an accomplice of Fantômas. *Third part*: the skill of the policeman Juve reveals that a VIP (sometimes a monarch) is none other than Fantômas in disguise. *Fourth part*: a fantastic chase leads Fantômas, pursued, and the policeman, pursuing, to various different places on the earth: Niagara falls, on the pipes of a locomotive, the outskirts of Grenoble, Cherbourg harbour, the outside of the Eiffel Tower, the bottom of the Atlantic, etc. In these

cases the authors show themselves not only to be extremely competent in geography, well up on the guidebooks, and expert drivers, but acute and terrible poets, dramatic authors of unique ingeniousness.' Max Jacob, 'Ecrit pour la S.A.F.', *Soirées de Paris*, 26–7 (1914). Quoted from Angot, *Fantômas revient*, pp. 48–9.

22 'The style is deplorable, let us not mince words, it is deplorable ... They misuse the word "henceforth".' Ibid.

23 'Full of life and imagination, written any old how, but very picturesquely.' *Le Mercure de France*, 16 July 1914, quoted from Apollinaire, *Œuvres en prose complètes*, III, 215.

24 'A *cultivated* public wildly enthusiastic about the adventures of the policeman Juve, of the journalist Fandor, of Lady Beltham and so on.' Ibid. (emphasis added).

25 But see Robin Walz, 'Serial Killings: Fantômas, Feuillade and the Mass-Culture Genealogy of Surrealism', *The Velvet Light Trap: A Critical Journal of Film and Television*, 37 (March 1996), 51–7, for an account of the importance of the films to the Surrealists, in particular of how Feuillade seizes upon and develops the crucial theme of identity displacement.

26 'These are even, I think, the only books I have read properly, and I have had the pleasure of meeting many right-minded people who shared this taste with me.' Apollinaire, *Œuvres en prose complètes*, II (1991), 974.

27 'Listen – Quiet please! – / to the melancholy rhymes / of nameless crimes / of torture and misdeeds, / all unpunished, alas! / done by wicked Fantômas! // Lady Beltham, his mistress / watched him do in her Lord / because they had been caught / in the midst of their caress. / He sank the liner Lancaster / down to Davy Jones's locker.' Robert Desnos, *Fortunes* (Paris: Editions de la N.R.F., 1942), pp. 105–13.

28 'The balance between two identical and opposing forces'; 'the enigmatic and tragic allegory of love, of its passions and its disorders'; 'humanity prey to hunger, vice and diseases, but which triumphs at last the day when Juve and Fantômas are dead. They destroy the old order.' See Marie-Claire Dumas, *Robert Desnos ou l'exploration des limites* (Paris: Klincksieck, 1980), p. 228.

29 'Belief in the higher reality of certain forms of associations ... in the omnipotence of dream, in the disinterested play of thought'. Breton, 'Manifeste du surréalisme', in *Œuvres complètes*, p. 328.

30 See Yves Olivier-Martin, 'Géographie de "Fantômas": Paris', *Europe*, 590/591, 119–40.

31 First published in *Distances* (March 1928); quoted from Gablik, *Magritte*, pp. 61 and 48.

32 Roger Shattuck, *The Banquet Years: The Origins of the Avant-Garde in France, 1885 to World War One*, revised edn, (London: Jonathan Cape, 1969).

33 'Who was not my type.' Marcel Proust, *A la recherche du temps perdu*, ed. Pierre Clarac and André Ferré (Paris: Gallimard, 1954), I, 382.

34 'For a long time I have been seeking refuge from poetic or realist literature in those books where the author is unaware that poetry and truth are haunting him, that they are lifting him beyond himself, and what's more above the scorn that he displays for a genre that he imagines unworthy of his pen; at the same time he is deceiving himself when he aims much

higher. Apollinaire and I had proof of this when the authors of *Fantômas*, astounded by our enthusiasm, informed us that they were capable of writing works that were much less naive. But these less naive works were disconcertingly naive. They showed us once again the primacy of a genius like that of childhood, a genius that those ill-educated persons from the college of the Muses confuse with clumsiness and chance.' Cocteau, 'Préface', quoted in Angot, *Fantômas revient*, p. 57.

35 Desnos, 'Imagerie moderne', p. 458.

36 'I wanted to write a *Life of Fantômas*.... I have read the thirty-two volumes four times; the fifth time I stopped at Volume 24'. Raymond Queneau, *Bâtons, chiffres et lettres* (Paris: Gallimard, 1965), p. 259.

37 Jean-Paul Colin, 'Les Verbes de la Mort dans *Fantômas*', *Europe*, 590/591, 96–104.

14
Bleeding the Thriller: Alain Robbe-Grillet's Intertextual Crimes

Jonathan C. Brown

> Je rêvais volontiers au massacre de mes camarades de classe... les corps gracieux aux jolis visages tendres avaient droit à de longs supplices, liés aux troncs des marronniers dans la cour de récréation.
>
> (Alain Robbe-Grillet)[1]

If self-conscious textual experimentation distances Robbe-Grillet from more orthodox crime writers, his taste for sadoerotic murder and his obsession with form place him among the most challenging practitioners of the genre. This essay argues that crime in his work is both the absorbing project of a complex subjectivity and a manipulable, mass-produced stereotype; further, that detection is both the search for a perpetrator and a metaphor for the process of interpretation associated with the act of reading within the instability of language. While Robbe-Grillet demonstrates that the textual representation of crime creates myths of a rational, unified self, he is dependent upon such representations for his subjective indulgence in the violence of sado-erotic fantasy. An examination of crime elements in certain key texts will illustrate both his formal subversion of generic stereotypes and the inevitability of narrative participation in the *mise-en-scène* of desire where what is prohibited is always present. Elements from crime fiction within self-reflexive, pluralistic texts form a network which reflects a self in process, both fixed and liberated at the limits of reading and rationality.

Perhaps Todorov is too extreme in his claim that 'the whodunit par excellence is not the one which transgresses the rules of the genre, but the one which conforms to them.' His statement relies on too clear a division between what he terms, in token inverted commas, '"high" art and "popular" art', and seems to ignore the promiscuous interchange between generic and non-generic forms which defines much postmodern literature.[2] Nevertheless even if a certain amount of rule-breaking is possible, it is in the nature of generic fiction to risk repetition as certain elements must be present for it to be defined as such. If innovation is not attempted, the genre dies, but if such innovation is pushed so far that crucial elements are discarded then the text may be excommunicated. The genre contains its mobility, and therefore to discuss the relation between the overtly experimental work of Alain Robbe-Grillet and genre writing runs the risk of reinforcing a negative view of crime texts.

An exploration of the links between Robbe-Grillet's work and crime writing is primarily justified by the prevalence of crime and investigation in his work. Robbe-Grillet's admission in his autobiography of a deep-rooted personal taste for sexual violence is played out in increasingly explicit fictional narratives which are often structured around generic crime plots. The significance of his use of such plots lies in their reflection of the artist's own thematic preoccupations but it also seeks to highlight the stereotypical nature of certain aspects of crime fiction. Nonetheless, as this essay will illustrate, the manipulation of generic stereotypes aims not to reinforce Todorov's opposition but rather to dissolve such a hierarchy by demonstrating the key rôle played by crime texts in the generation both of experimental texts, and of a fragmented and mobile subject.

Les Gommes (1953) bears closest resemblance to the detective novel in its apparent adoption of the crime, investigation, resolution pattern which characterises Golden Age and much hard-boiled detective fiction. Wallas, a special agent for the 'Bureau des Enquêtes', investigates the killing of a prominent local figure, Daniel Dupont, by an anarchist criminal gang. After a series of fruitless interrogations and false trails, Wallas returns to the scene and is interrupted by none other than Daniel Dupont himself, who has in fact only been injured. It is here that the fragile scaffolding of the generic structure collapses: acting on reflex, Wallas shoots Dupont dead, thereby committing the crime he has been investigating.

This summary conveys something of the playful nature of *Les Gommes*. Its intrinsically ludic quality coupled with a rejection of realism characterise Robbe-Grillet's work and that of other practitioners of the Nouveau Roman. The awareness of the artificiality of writing – its status as fiction rather than as reflection of 'truth' – in some respects aligns explicitly experimental work with crime fiction and explains

Robbe-Grillet's kinship with the genre. In many ways crime fiction is playful as one of its main features is the ludic use of mystery. As Dennis Porter rightly suggests, irony is central to detective novels in the form of the 'surprise of crime' which is expressed through the inevitable emergence from work to work of the most unlikely suspect, the most unlikely detective figure (seen in the unsuitability of characters like Miss Marple) and the eccentricity of the setting, all of which set up the 'ambience of play'.[3]

If certain crime writers do not question stereotypes and thereby encourage a perception of crime writing as formulaic, the generic nature of many aspects of crime fiction is itself a spur to others to experiment with subject matter and theme. The conscious use of generic plot structures paradoxically increases the possibilities for experimentation, reflected in the very diverse subjects chosen as settings for contemporary mysteries, such as the plight of Greenlanders in Peter Hoeg's *Miss Smilla's Feeling for Snow* or the Holocaust in Philip Kerr's *Berlin Noir* series. The game of detection may be endlessly replayed. The reader detects on two levels, not just empathising with (or trying to beat) the hero or heroine as he or she solves the crime, but also tracing tell-tale clues from one book to the next and measuring the success of the writer by his/her ability to perpetrate an age-old crime in a cunning new way.

And like the work of Robbe-Grillet which forefronts the partiality of its vision, the crime novel has not claimed to represent 'reality'. Criminals in Sherlock Holmes stories are not members of the dangerous classes who genuinely threatened bourgeois London at Doyle's time of writing but instead are 'respectable people gone wrong, turned aside from their proper rôles'. In Chandler, it is Marlowe's 'controlling and isolative feelings' which dominate the physical environment rather than the social realities of the time.[4] A concern with realism is optional in crime writing, illustrated perhaps most clearly by the existence of a separate genre called 'True' Crime.

There are however many ironic, metatextual strategies at work in Robbe-Grillet's texts which erode the myths perpetrated by much crime writing. *Les Gommes* questions and negates the hermeneutic process which pervades crime fiction and is most evident in Golden Age detective stories. Dupin and Holmes embody a belief in rational deduction as they master the mysteries of the world through the exploitation of empirical thought processes. This in turn implies that events are linked in an unaccidental chain and that the universe is indisputably coherent. Robbe-Grillet relies on the reader's generic competence in order to deconstruct such assumptions. The most evident inversion effected by *Les Gommes* is Wallas's rôle as both sleuth and murderer. The failure of events to form a chain of coherence culminates in this most catastrophic of generic implosions.

Yet the implication that the world will not and cannot surrender its mysteries to the investigative process is played out on many levels of the text. Employing a detailed geometric precision to describe what is insignificant to the plot, Robbe-Grillet places emphasis on the bizarre nature of physical reality and opaque nature of characters. The narrative dwells on inconsequential moments and does not explain certain sub-plots at all, such as the murder of local girl Pauline, 'morte d'étrange façon', which in turn disrupts the cause-and-effect machinery.[5] The authority, superiority and isolation of Dupin and Holmes is undermined in *Les Gommes* by Wallas's constant confusion and by the fact that his thoughts and actions uncannily echo those of other characters. Rather than controlling interrogations, Wallas himself seems suspect as his undercover rôle forces him to invent lies in which he eventually becomes lost.

Golden Age detective fiction and certain modern variants such as the police procedurals of Ed McBain have been seen rightly by commentators as characterised by a central paradox which perpetrates a significant deception.[6] These texts promote a myth of the superiority of the individual, exemplified in his/her ability to solve cases and by the constant stress on the act of individual betrayal as the only real crime, but in order to do this they must conceal the transgression at the heart of the individual's psychological fantasies and thereby misrepresent the nature of individuality. In such texts the 'guilty' party is the one who fails to conform to social codes; this denies the nonconformism which defines individual freedom. Robbe-Grillet's work presents a view of the self as neither fixed nor unified due to constant disruption caused by unconscious impulses and repressive forces.

Admittedly hard-boiled crime writers from Hammett to Edward Bunker and James Ellroy have consistently challenged such a simplistic resolution between the individual and society. Scott Turow's protagonists possess neither self-knowledge nor a superior moral sense. *Presumed Innocent* is not so much an investigation of a crime as an exploration of the depths of Rusty Sabich's unresolved sexual guilt. Michael Dibdin's Zen novels maintain a double focus on the solution of the mystery and on the manipulation of Zen himself by the corrupt Italian political system. James Ellroy's 'crime memoir', *My Dark Places*, is an obsessive meditation on the writer's own identity through the figure of his absent mother, rather than a coherent investigation of her murder. The book reveals the perpetual contradictions which underlie the writer's and his mother's self-images.

Many mysteries in the Golden Age tradition however propose a model for an unproblematic textual representation of a knowable 'reality' and this has tended to pervade crime writing. The text which promotes a myth of resolution between individual and society does this by perpetuating semantic stability. As Todorov points out with regard to Golden Age detective fiction, one story hides another only to reveal the latter's 'truth'. Misrepresentation is simulated so that it can be corrected and in order that the illusion of a transparency between version and event, between language and 'truth' can be reinstated. If Eco is correct in his assertion that 'clues are seldom coded, and their interpretation is frequently a matter of complex inference rather than of sign-function recognition', nonetheless Moretti is equally correct in his statement that the criminal's guilt stems from his/her creation of semantic ambiguity which the detective corrects by reimposing a univocal link.[7] It is no accident that the investigation is often written in a style which is 'perfectly transparent, imperceptible ... simple, clear, direct',[8] and which relays facts rooted in a rigidly planned timetable: such stylistic features exemplify the fantasy perpetrated by the very structure, namely, that univocal, reliable readings of a knowable 'reality' are possible. While *Les Gommes*'s detective must sacrifice his textual meaning before locating his corpse, the body of much crime fiction begins as a site of illegibility and is then reassuringly recuperated within stable semantic structures which eventually spell out the name of the perpetrator, in serial killer texts by way of a 'profile' – a hypothetical biography based on a tell-tale 'signature'. In this way, death's aporia is neutralised and incorporated within a teleologically coherent narrative.

Robbe-Grillet's dramatisation of a failure of reading through the detective's failure to interpret is developed further in the film and cine-novel *Glissements progressifs du plaisir* (1974) in which pieces of evidence play a complex structural rôle. Instead of revealing the identity of the killer, they form a sadoerotic mythology of their own which undermines the rational investigation undertaken by policeman, prosecutor and priest. Alice is suspected of the violent stabbing of her lesbian lover while they were in bed together. She is presented with evidence of her guilt in the form of a high-heeled shoe worn by both her and the victim, and a prie-dieu and broken bottle with which Alice has performed blasphemous, sadomasochistic rôle-play. As the significance of each item within the crime plot is sought, these clues generate new stories narrated by Alice and performed in either fantasy images or flash-backs, which in turn contradict events known so far and absorb the interrogators themselves in sadoerotic narratives tangential to the crime. The priest eventually gives in to his obsession with lesbianism evoked by Alice's fantasies; the prosecutor becomes insane and takes to his bed. Rational explanations dissolve, like the father's attempt to recover his daughter from kidnappers in Robbe-Grillet's film *Le Jeu avec le feu* (1975) which is perverted by his own incestuous fantasies. In this way clues which would conventionally contribute towards semantic unity instead produce a multiplicity of meanings and destroy notions of an 'individual' guilt which may be contained and controlled.

As may now be clear, Robbe-Grillet derives drama not only from the often gory spectacle of violence, but also from the frustration of expectations and deliberate misuse of narrative conventions. Suspense is generated not only from the anticipation of crime but also from uncertainty about the source of narration and the status of information which in turn unsettles the reader's attempts to define events as real, imagined or remembered. Despite an often self-advertising artificiality, the reader or film-viewer cannot escape complicity in fantasies of crime which are never resolved by the reimposition of moral, social or even textual order. This also indicates the pervasive sense that criminal fantasies are universal and so cannot be dispelled by the hierarchical structure of much crime fiction.

Le Voyeur (1955) shifts the focus from the process of investigation to the activities of a psychopath and in this sense is less a 'detective novel' than a 'crime novel'.[9] Mathias, a watch-salesman, visits the island where he was born in order to tout his wares. After a failed attempt at selling watches his real preoccupation emerges when the naked corpse of a young shepherdess is washed up on the shore. Stumblingly the narrative reveals that Mathias may have been involved in the crime which involves bondage and torture. The crime is then forgotten as Mathias continues on his travels.

If the parodic treatment of the crime genre evident in *Les Gommes* is absent from *Le Voyeur*, both texts are aligned in their investment of violence in the very act of narrating. As Ian Bell maintains, a novelist such as Patricia Highsmith successfully erodes the distinction between rational and irrational realms by describing acts of psychopathic violence with lucidity and by refusing moral judgement; Robbe-Grillet conveys such criminal 'insanity' through the external narrator's apparent inability to de-cypher thoughts and events. This captures the violence of crime through a metatextual attack on the process of representation itself.[10] In their very fragmentation, the faltering description of Mathias's behaviour, and the recurrence of obsessive, sadistic but displaced images, manage to reflect the psychopathic mind with its 'inordinate amount of instinctual aggression and ... absence of an object relational capacity to bond' which leads to 'a fundamental disidentification with humanity'.[11]

The reader only learns about the crime when Jacqueline Leduc is washed up on the shore on page 174, a long time after the act of murder takes place, supposedly on the blank page 88. Yet even when the crime has been admitted, the text offers very little information, dropping hints such as references to cigarettes which Mathias left at the scene which may have been used to burn the victim: 'C'est alors qu'il pensa, pour la première fois, aux trois bouts de cigarettes oubliés sur la falaise.' In flashback, or in his inventive imagination, Mathias remembers the crime: 'Il voyait la petite bergère étendue à ses pieds, qui se tordait faiblement de droite et de gauche. Il lui avait enfoncé sa chemise roulée en boule dans la bouche, pour l'empêcher de hurler.'[12] However the narrator's failure to inform the reader earlier on removes authority from these statements and renders the entire account somewhat suspect. The absence of police investigation and hazy normality at the end of the text leave the reader uncertain as to whether a crime has really taken place at all.

The absence of fixed chronology and topography mean that the reader is excluded from an attempt at investigation. Unlike the objectified account of time and place which begins in the Golden Age detective story and runs throughout modern mystery writing, Mathias inhabits a subjective time which cannot be calibrated. The crime cannot therefore be retrospectively plotted as the criminal's misrepresentation of himself has corroded external narration. Suspense is generated not just by the gradual revelation of violent confrontation but also by uncertainty about whether events are happening in reality, in imagination, in memory or in dream. The conflict between readings mirrors the psychological drama inside Mathias's head so that the reader too is trapped within an unstable process with no fixed co-ordinates. The dissolution of objectivity of the external narrator implies that this sole source of authority within the text is also covering up the crime, conveying an unsettling sense that the sadosexual fantasy has permeated all attempts to control it.

Psychological instability mirrored in a fragmentary narrative also characterises *Un Régicide* (1949) which, while closer to a magical fable than either detective story or crime novel, focuses on an assassination attempt, and thereby relies on and undermines a generic plot. In contrast to mainstream stories of assassination, such as Oliver Stone's film *JFK* or Forsyth's novel *The Day of the Jackal* which play on contemporary conspiracy theories, Robbe-Grillet's first novel depicts an assassination which has no explicitly psychological or political motive. The instruction to assassinate the king is first given to Boris by an anonymous voice in a crowd. The second command is generated by Freudian word association. In an apparently irrelevant sub-plot, a student called 'Red' has died, killed either by police or as a result of a 'crime passionnel'. When visiting his grave, Boris reads the words 'Ci-gît Red' ('Here lies Red') which his mind reorganises immediately into the near anagram, 'Regicide!', inspiring his fatal act.[13] The assassination itself takes place twice, the first attempt leaving the king unharmed and the second lapsing into comedy as Boris stabs the king and is immediately deafened by gunshots, thus unable to tell whether he has succeeded or not. Despite bloodstains on the knife, the media only later admit that the king has fallen ill. Boris is arrested but the case against him dissolves.

In his adoption of a popular paradigm, Robbe-Grillet reveals the ritualistic nature of narratives of political assassination and suggests that they fulfil an unconscious desire both to celebrate and destroy symbols of authority. While a more naturalistic story of assassination would enable the reader to channel such impulses into a containing structure but ultimately denies him/her anything other than escapism, Robbe-Grillet places the origin of the command to assassinate in a process of verbal slippage, and thereby lends language itself both an intrinsic criminal violence and a potentially revolutionary power. A violent instability is perpetrated by the verbal structures of the narrative itself as events generate multiple meanings. This in turn renders the process of reading a participation in a crime against textual convention and authority.

If *Le Voyeur* achieves a perversely shocking realism through its very reluctance to narrate, other texts by Robbe-Grillet are characterised by a self-reflexivity which undercuts illusions of realism. Yet while self-consciousness highlights the textual status of what is presented and thereby draws attention to the generic nature of elements drawn from the crime novel, the novels' very reliance on generic stereotypes implies a deep-rooted subjective investment in them, and a sense that narrator and writer are exploiting such stereotypes as sites of their desire.

Projet pour une révolution à New York (1970) conjures up a dreamlike New York which is as much a psychological and verbal labyrinth as a city. The text does not adopt a generic model for its overall scheme but instead exploits a series of fragments from crime fiction such as undercover agents, break-ins, rapes and phoney doctors. The text celebrates sadistic torture through a series of unrestrained fantasies while also interrogating itself on its plausibility and self-consciously highlighting the generic nature of its contents. The investigations and pursuits in this novel culminate in the lengthy torture scene in which a prostitute, who may also be a clothes-dummy, straddles an upended saw and is slowly torn apart. This scene is also a staged game taking place on a giant chess-board which cannot be explained logically other than as a subjective fantasy of the narrator.[14]

Sadoerotic crime in *Projet pour une révolution à New York* is also a violation of narrative structures. Chronology, topography and the identities of characters and narrators are so consistently confused that perpetrator and victim become virtually interchangeable. Verbal games fragment the realism of crime but become indissociable from it, so that the violent scene thematizes the workings of a self-reflexive narrative which both creates and destroys itself as it proceeds towards the fulfilment of its perverse desire. The word itself is put on the rack and either forced into giving a stream of contradictory meanings – like a victim who will say anything to make the torturer stop – or spliced into simple letters, as names are reduced to initials such as JR, W and MAG. The act of torture is thematically linked to the act of writing through the administering of the truth drug ('sérum de vérité') by the sinister Dr Morgan, a corrupt doctor in the style of Chandler's Dr Sonderberg of *Farewell My Lovely*, Dr Loring of *The Long Goodbye* or Robert B. Parker's Dr Croft in *God Save the Child*. The truth drug becomes a catalyst of textual production as the victim speaks so much that s/he takes over narration of the text.

The intertextual relationship between Robbe-Grillet's texts and crime fiction reveals itself not only in the ironic assemblage of generic stereotypes but also in the juxtaposition of such elements with intertexts from other genres and media. Intertextual cross-fertilisation in turn questions and dissolves boundaries between genres. Elements from crime fiction collide with narratives from respected texts such as the Bible and works by authors such as Goethe in a process designed to erode the individuality and perceived 'superiority' of such texts.

Robert Brock has explored the extent to which *Les Gommes* and *Le Voyeur* borrow certain plot elements from Graham Greene's books *A Gun for Sale* and *Brighton Rock*, the first of which Greene subtitled 'an entertainment' so that it would not be confused with his 'proper novels', the second of which began as a detective story.[15] While Robbe-Grillet's use of these intertexts is playful and destructive (Raven, Greene's assassin in *A Gun for Sale*, becomes in Robbe-Grillet's text an amalgamation of detective and hit-man), the collision between detective form and Sophocles' *Oedipus the King* in *Les Gommes* generates the most dramatic intertextual collision. References to the Oedipus myth replace criminal traces as it is implied that Dupont may be Wallas's father. The detective's association with 'Oedipe' is enhanced by his hobby of crumbling erasers in his pocket which are inscribed with the initials '-di-' and by the depictions of Thebes in the stationery shop window. The reader is invited to construct the intertext of *Oedipus the King* through references such as the drunkard's riddle which begins on page 17 as a fragment, 'Quel est l'animal qui, le matin…', and culminates in the bluntly incriminating 'Quel est l'animal qui est parricide le matin, inceste à midi et aveugle le soir?'[16]

On the one hand Robbe-Grillet is highlighting the similarity between the detective form and the story of Oedipus which has also been suggested by psychoanalysts.[17] The intertextual superimposition simultaneously questions the legitimacy of each narrative type as Wallas's rôle as Oedipus undermines and ridicules his rôle as a detective. The circularity of the Oedipus story with its inevitable, tragic ending collides violently with the supposedly rational, linear quest of a detective who would conventionally form a watertight account of events and remain authoritatively outside them. Far from reproducing the assumptions of the generic models, *Les Gommes* uses one to undermine the naiveté of the other, making intertextual collision a significant act of metafictional parricide.

Robbe-Grillet's films frame an eclectic and challenging range of intertexts and therefore generate striking generic disruption. The 1983 film *La Belle Captive* begins in the style of hard-boiled detective fiction as Walter, the hero, searches for a missing girl who may or may not be dead. Soon however the intertext of Goethe's poem 'Die Braut von Korinth' ('The Bride of Corinth') reorientates the story as the woman Marie-Ange becomes a vampire, sucking Walter's blood, both dead and alive. In Goethe's poem a woman murdered by her own parents becomes a vampire and wins a young man's heart; by the time he has realised she is one of the 'undead', he too has been bitten.

References to detective fiction are then undercut by intertextual allusions to Edouard Manet's painting *The Execution of Maximillian* and Magritte's *Le Mois des vendanges*. These pictorial images appear in the film and then become generators for the text as Walter is pursued by a firing-squad for his sexual activity with the (possibly dead) girl, as in the Manet picture. Magritte's vision becomes an animated episode in the film generated from Walter's tortured mind while he is held captive in a clinic run by another phoney doctor.

Cross-fertilisation also occurs in the film *Trans-Europ-Express* (1966), a film about the making of a thriller. Allusions to Hitchcock's films dynamically erode Biblical references effecting thematic implosion and generic evacuation. Intertextual reference to the Bible is most explicitly made via the names of the protagonists, Mathieu, Marc, Lucette and Jean, which uncannily echo the names of Christ's apostles, Matthew, Mark, Luke and John. Whereas the narrators of The Gospel in The New Testament are driven by a common purpose, each in turn recounting 'The Coming of Christ', in Robbe-Grillet's film no such correspondence between narratives is attained, as Jean, Marc and Lucette all disagree over how the story should proceed. They eventually lose track of Elias, the hero of their projected film, as he pursues a narrative of his own by strangling a prostitute. The collapse of authority among the narrators works against the assertion of absolute truth in the biblical intertext evoked in the film. The conflict between narrators suggests an instability and provisionality within the Bible itself which is indirectly compared to Robbe-Grillet's network of self-cancelling histories.

Moreover references to the biblical intertext collide with allusions to Hitchcock's 1951 adaptation of Patricia Highsmith's crime novel, *Strangers on a Train*. In the intertext, Bruno Anthony performs his killing by strangulation and then pressures Guy Haines to follow the plot they had both set up in a text overloaded with motive. In Robbe-Grillet's film Elias's strangulation of the prostitute has no place whatsoever in the narrative of the authors within the film. And whereas in *Strangers on a Train* Bruno also plays a part in the events he narrates, thereby implying causality and coherence between narration and action, in Robbe-Grillet's film an ironic and often misleading distance is maintained between Jean (one of the authors within the film) and Elias's narratives. While *Strangers on a Train* pits Bruno's narration (within the film) against the social and moral order and culminates in a reinstatement of this order through the death of Bruno when the forces of order catch up with him, *Trans-Europ-Express* pits narration against narration and culminates in dissolution when Elias is killed at the end of the film only to come back to life minutes later in the last sequence. The intertextual network brings the detective or thriller form into collision with biblical references. Robbe-Grillet's anti-hero Elias is Christ when situated within the biblical model, but becomes a stereotypical psychopath in the Hitchcock context. This ironic intertextual reference corrodes the unity of these personae and undermines the authority of both intertexts.

If crime fiction plays a central rôle in the intertextual disruption effected by Robbe-Grillet's work this is partly due to a desire to celebrate the 'ambience of play' of such work but it also alerts the reader to the naiveté underlying certain elements of the genre. Crime fiction depends on coherence and logic, on the authority of the narrator or hero, and on an objectified account of time and place which Robbe-Grillet's work attempts to invalidate. Yet the elements from crime fiction in his work also comprise a personal mythology. Crime invades all levels of a textual process which both disrupts narrative order and implies that criminal fantasy is pervasive. The self-advertisingly false nature of crime elements, which also structure dream and fantasy, indicates their double role as both mass-produced stereotypes and images invested with subjective meaning. Their violence is not just the sado-sexual torture they depict but also an erosion of the notion of a stable self which is reduced to expressing its prohibited desires through images from sadomasochistic pornography and crime fiction. The personal mythology generated from such generic stereotypes is both fractured and impersonal; the textual combination of elements from crime fiction reveals that the artist's language is itself derivative and inauthentic.

Yet the collage of generic elements from both crime fiction and other texts within fragmented narrative structures also reveals an awareness of a dependence on ready-mades which, far from restricting the artistic process, on the contrary enables a freedom to work with them in a challenging and mobile way. If generic components are deformed via their insertion within unstable texts, they are also celebrated and re-affirmed. The intertextual collision between elements from crime fiction and more traditionally respected texts erodes the hierarchy between 'high' art and 'popular' fiction but also highlights the importance of both in the constitution of any textual representation of subjectivity.

Notes

1 'I took pleasure in imagining the massacre of my school-mates. ... Those with graceful bodies and pretty, tender features would be singled out for lengthy torture while tied to chestnut trees in the playground.' From the first volume of Robbe-Grillet's autobiography, *Le Miroir qui revient* (Paris: Les Editions de Minuit, 1984), p. 180.

2 Tzevan Todorov, 'The Typology of Detective Fiction', in David Lodge (ed.), *Modern Criticism and Theory: A Reader* (London: Longman, 1988), pp. 158–65.
3 Dennis Porter, 'The Language of Detection', in Tony Bennett (ed.), *Popular Fiction: Technology, Ideology, Production, Reading* (London: Routledge, 1990), pp. 81–93 and 86.
4 Stephen Knight, *Form and Ideology in Crime Fiction* (London: Macmillan, 1980), pp. 90 and 145.
5 'Pauline, who died in a strange manner', *Les Gommes* (Paris: Les Editions de Minuit, 1953), p. 15.
6 The most sophisticated explanation of the contradiction between the myth of the unified individual and its relation to society is provided by Stephen Knight throughout his book *Form and Ideology in Crime Fiction.*
7 Umberto Eco, *A Theory of Semiotics* (London: Macmillan, 1977), p. 224; Franco Moretti, 'Clues', in *Popular Fiction: Technology, Ideology, Production, Reading,* pp. 238–51 and 249.
8 Todorov, 'The Typology of Detective Fiction', p. 161.
9 For helpful guidelines on the difference between these types, see Julian Symons, *Bloody Murder: From the Detective Story to the Crime Novel: A History* (London: Pan, 1994), pp. 201–3.
10 For a revealing study of innovation in Highsmith, see Ian A. Bell, 'Irony and Justice in Patricia Highsmith', in Ian A. Bell and Graham Daldry (eds), *Watching the Detectives: Essays on Crime Fiction* (London: Macmillan, 1990), pp. 2–6.
11 J. Reid Meloy, *The Psychopathic Mind: Origins, Dynamics, and Treatment* (London: Jason Aronson, 1988), pp. 5 and 41–7.
12 'It was only then that he remembered the three cigarette ends which he had left on the cliff-top.' 'He could see the little shepherd girl stretched out before him, twisting her body feebly from left to right. He had stuffed the rolled-up blouse into her mouth to stop her crying out.' *Le Voyeur* (Paris: Les Editions de Minuit, 1955), pp. 178 and 179.
13 *Un Régicide* (Paris: Les Editions de Minuit, 1978), p. 58.
14 *Projet pour une révolution à New York* (Paris: Les Editions de Minuit, 1970), pp. 176–86.
15 Robert R. Brock, *Lire, Enfin, Robbe-Grillet* (New York: Peter Lang, 1991), pp. 4 and 39.
16 'Which animal is patricidal in the morning, incestuous at noon and blind in the evening?', *Les Gommes,* p. 234.
17 For a useful summary of both Geraldine Pederson-Krag's 'Detective Stories and the Primal Scene' and the development and revision of her ideas by Charles Rycroft, see Symons, *Bloody Murder,* pp. 19–20.

15
Oedipus Express: Trains, Trauma and Detective Fiction

Laura Marcus

I

It is possible to buy a short thriller or detective story from vending machines in Paris railway and underground stations. A spokeswoman for the publisher, Editions de la Voute, said in early 1997, 'the idea is just a rethink of the success of station and airport novels. Since commuter journeys tend to be short, we decided that none of our books should take more than one hour to read.' The novel offered is changed every week, alternating between a new book and a reprint. The first offering in the Metro-Police series, as it is called, was Gerard Delteil's *Le nouveau Crime de L'Orient Express*, a new version of Agatha Christie's famous train-crime novel.

Walter Benjamin noted the elective affinity between the railway journey and detective fiction in an article published in 1930 and entitled 'Kriminalromane, auf Reisen' ('Detective Novels, on Journeys').[1] In this short piece, Benjamin, who, like a number of the theorists associated with the Frankfurt School, was fascinated by detective fiction and indeed wrote a synopsis of a detective novel,[2] observed that railway travellers prefer to buy their reading material from station bookstalls, opting for the contingency of what they find there rather than for the certainties of what they might bring from home:

> less out of pleasure in reading than in the dim feeling that they are doing something which pleases the gods of the railway. He [the traveller] knows that the coins which he offers up to this sacrificial column recommend him to the protection of the boiler god which glows through the night, the smoke naiads which play above the train, and the demon who is master over all lullabies. He knows all

> of them from dreams, but he also knows the succession of mythical trials and dangers which has presented itself to the Zeitgeist as a 'railway journey', and the unforeseeable flow of spatio-temporal thresholds over which it moves, starting with the famous 'too late' of the person left behind, the original image of everything missed, through to the loneliness of the compartment, the fear of missing the connection and the grey of the unknown station into which he is riding. Unsuspectingly, he feels himself drawn into a giants' kingdom and recognises himself to be the mute witness of the struggle between the gods of the railway and the station.
>
> Simila similibus. The anaesthetising of one fear by the other is his salvation. Between the freshly separated pages of the detective novel he seeks the leisured, almost virginal anxieties which can help him over the archaic anxieties of the journey.[3]

As he lists the detective novelists the traveller might most wish to read, Benjamin's main criterion appears to be the fittingness of the detective story's shape and structure to the rhythms of the journey. One might also turn, he suggests, to the stories of Gaston Leroux (author of *The Phantom of the Opera*) for an 'understanding of the future into which one is travelling, for the unresolved riddles left behind', or to Anna Katherine Green (author of *The Leavenworth Case*), whose 'short stories are just the same length as the Gotthard tunnel and [whose] great novels *Behind Locked Doors, In the House Next Door,* bloom in the purple-shaded coupé light like night violets.'

'So much', Benjamin concludes, 'for what reading offers the traveller.' But what does travelling offer the reader? 'When else', he asks,

> is [the reader] so involved in reading and can feel his hero's existence so certainly combined with his own? Is his body not the weaver's shuttle which follows the rhythm of the wheels in tirelessly shooting through the paper? ... People didn't read in post-coaches and they don't read in cars. Travel reading is as closely linked to railway travel as is the time spent in stations. (p. 382)

In this chapter, I take up the question of train travel, narrative and reading, with a particular focus on the affinities between train travel and detective fiction, which are multiple, as Benjamin suggests. From Benjamin's arguments we can begin to draw out the relationship between railway travel and the 'railway novel', which arises as a genre in the mid-nineteenth century to meet, or perhaps to shape, the needs of travellers. This is closely linked to the reflexive structure of reading about a train journey while journeying on a train: Benjamin instances Holmes and Watson, in 'the strange familiarity ['das Unheimlich-Heimliche'] of a dusty second class coupé, both sunk in silence as passengers, one behind the windshield of a newspaper, the other behind a curtain of smoke clouds' (p. 382). This line branches off into the many and various representations of trains and train journeys in detective fiction, some of which are discussed below. There is also, in Benjamin's account, the relationship between the movement of the train and the movement of the text from one place to another, departures and arrivals, the rhythms of the journey and the rhythms of the narrative. This rhythm combines with the physical sensations experienced by the train traveller, the movement of the train felt through the body, producing, in Benjamin's words, not only 'a shudder of tension' but a very particular and very intense kind of identification with the trajectory of the hero's destinies and destinations. Underlying these relationships (although this concept is not present in Benjamin's very brief discussion) is the simultaneity of production and consumption in railway travel, whereby the industrial machinery 'produces' the traveller as both consumer and commodity. Conjoining them is the observance of what Mark Seltzer in his recent work on serial killing, with reference to the nexus of 'body' and 'machine' in Naturalist fiction, has described as a 'logic of equivalence'.[4]

The logic of equivalence (journey/book, body/machine) exists, however, in relation to, and in tension with, the logic of belatedness, the 'missing' of points in space and time on which Benjamin also insists. In his account, the logic of equivalence, of synchronicity and simultaneity, serves to suppress or mask the anxiety of 'everything missed'. Yet the logic of equivalence can also be the logic of violence and of murderous repetition. If, as Seltzer suggests, the logic of equivalence is the driving force behind the Naturalist novel of the end of the nineteenth century, the logic of belatedness may be more characteristic of modernism, with, in Thomas Hardy's phrase, its 'ache', its awareness of things 'passing', and in Freudian terms its dynamics of deferral and *Nachträglichkeit*. The railroad, as Wolfgang Schivelbusch suggests in his brilliant study *The Railway Journey*, 'knows only points of departure and destination.'[5] The spatial dimensions of the present are thus effaced: there is only the past and the future.

Benjamin writes of 'the struggle between the gods of the railway and the station', thus calling attention to a complexity that, as I argue later, is crucial to detective fiction in its relation to trains. 'The railway' is made up of a number of different and disparate structures and mechanisms – engines, carriages, tracks, viaducts, embankments, bridges, tunnels, stations – the links and spaces between which detective fiction and thrillers construct plot and narrative movement. In his essay 'Railway Navigation and Incarceration', Michel de Certeau introduces a further element of the railway, the windowpane, and a logic we can add to the logics of equivalence and of belatedness, 'an imperative of separation'.[6] The traveller is 'immobile inside the train, seeing immobile things slip by', less a reader than a character or 'type':

> Every being is placed there like a piece of printer's type arranged in military order. This order, an organizational system, the quietude of a certain reason, is the condition of both a railway car's and a text's movement from one place to another. ... Between the immobility of the inside and that of the outside a certain *quid pro quo* is introduced, a slender blade that inverts their stability. The chiasm is produced by the windowpane and the rail. ... The windowpane is what allows us to see, and the rail, what allows us to *move through*. These are two complementary modes of separation. The first creates the spectator's distance: You shall not touch; the more you see, the less you hold. ... The second inscribes, indefinitely, the injunction to pass on ... an imperative of separation which obliges one to pay for an abstract ocular domination of space by leaving behind any proper place, by losing one's footing. (pp. 111–12)

De Certeau argues that 'it is the silence of these things put at a distance, behind the windowpane, which, from a great distance, makes our memories speak or draws out of the shadows the dreams of our secrets' (p. 112). Benjamin points to 'the succession of mythical trials and dangers which has presented itself to the Zeitgeist as a "railway journey" ';[7] de Certeau suggests the relationship between the quotidian or rationalised aspects of the train journey and the train as dream and dream-symbol. He also alludes to the 'railway journey' as an experience which blurs the boundaries between the imaginary and the real.

The paradox in de Certeau's account is that the passenger's location in an ordered system leaves him or her free or, perhaps, constrained to dream. His entire construction of the 'railway journey' is founded on the rhetorical structures of paradox and chiasmus. Thus he points to the inversion of reason/dream; inside/outside; mobility/immobility. At the close of his essay, de Certeau describes the incongruity of the train, the 'immobile machine', in 'the mobile world of the train station' (p. 114).

In the third section of this article, I explore the ways in which the three logics which I have identified here – the logics of equivalence, belatedness/missing, and separation/inversion – are worked through different kinds of crime and detective fiction. As a prelude to this discussion, I look in Section II at the terms of trains and trauma, and at the development in psychology and psychoanalysis of the logics which I subsequently trace through fictional forms. Thus Walter Benjamin's language of narrative identification, simultaneity and temporal 'missing' finds its counterpart in the psychological discourse of hysterical identification, 'neuromimesis', and the psychical deferral generated by the railway and its accidents.

II

In the cultural narratives of the railway in the nineteenth and early twentieth centuries, the categories of shock and trauma are central. Freud's 'Three Essays on the Theory of Sexuality' contain a striking example of the swerve from the anticipated account of the railway as a vehicle for smooth linear progress to a model which emphasizes 'mechanical agitation of the body'.[8] Although Freud's account describes a very specific trajectory – railway-travel arouses sexual impulses in the male child which are subsequently repressed and re-emerge as railway anxiety – the model of shock (here 'mechanical agitation') is not only inseparable from the discourses of the railway, but emerges at a number of historical junctures. These include the 1860s, when 'railway shock' was extensively conceptualized and when the discourse of fictional 'sensation' reached its height; the 1890s, when the nineteenth-century obsession with nerves and nervousness met the languages of degeneration; the 1920s, when Freud, Benjamin and others most substantially theorised the concept of 'shock'; and the present day, when we retrace a certain history of modernity as a history of shock, fragmentation, dispersal, heterogeneity.

The following text was published in 1884:

> The man, for the time being, becomes a part of the machine in which he has placed himself, being jarred by the self-same movement, and receiving impressions upon nerves of skin and muscle which are none the less real because they are unconsciously inflicted.[9]

These lines, as Schivelbusch has noted, are indicative of the concern in the last decades of the nineteenth century with the effects of rail travel on passengers and railway personnel. The terms of the quotation construct a striking logic of equivalence: man and machine are 'jarred by the self-same movement'. If the railway destroyed the mimetic relationship between pre-industrial transport and natural phenomena in its felt 'annihilation of time and space', mimesis is insistently returned in the equivalence between, and perceived interchangeability of, human and machine organisms.

Schivelbusch argues that the railway effected the most profound of revolutions in the nineteenth century, confronting the bourgeois traveller with the industrial process. The history of such conceptual and perceptual rupture is a history not only of progress but also of shock and trauma. These last terms are themselves reinvented, or come into being, to describe the effects of industrialised modernity and the disasters of the industrial 'accident' on the human organism. The anxieties attached to the railway are at least fourfold, although they may be temporally staged, one replacing (but not entirely effacing) the other: the initial difficulty of adjustment to mechanized travel and new temporal-spatial relations; the fear of the railway accident; anxiety attaching to the specific architecture of trains, especially their isolated compartments without corridors; concern over the effects of mechanized motion on the human nervous system, a concern which becomes increasingly insistent in the later nineteenth century in both medico-legal and psychological discourses. The human organism is both analogized as an engine or locomotive, and threatened by the machinery of the modern age, which depletes its energies and produces nervous exhaustion, or 'neurasthenia'.[10]

The 1860s saw a proliferation of medical literature on the manifestations of railway-related fear and nervous disorder. The vibrations to which travellers were subjected caused, it was believed, nervous, muscular and mental exhaustion. In 1892, Max Nordau wrote in his notorious work *Entartung*, or *Degeneration*: 'Even the little shocks of railway travelling, not perceived by the consciousness ... cost our brains wear and tear.' He added that the so-called 'railway neuroses' were

> exclusively a consequence of the present condition of civilized life ... the terms 'railway-spine' and 'railway-brain', which the English and American pathologists have given to certain states of these organs, show that they recognise them as due partly to the effects of railway accidents, partly to the constant vibrations undergone in railway travelling.[11]

The concept of 'railway spine' became prevalent in medico-legal thinking from the 1860s onwards, as increasing numbers of railway accident victims sought medical witness when claiming compensation from railway companies.

The most influential theorist of 'railway spine' was John Erichsen, a Professor of Surgery concerned with spinal injuries from accidents. They occur, he wrote, 'in none more frequently or with greater severity than in those which are sustained by persons who have been subjected to the violent shock of railway collision ... consequent on the extension of railway traffic.'[12] Erichsen's location of a set of symptoms with a pathological cause (degeneration of the spinal cord) was subsequently superseded by a focus on psychological factors, especially fright, in railway accidents and a concept of nervous shock, a traumatisation of the victim without there being discernible physical injury. Herbert Page, a surgeon for the London and Northwest Railway writing in the 1880s, focused on 'the profound impression upon the nervous system' in railway accidents, at times 'sufficient to produce shock, or even death itself.'[13] The perception of railway danger is certainly sustained (if not produced) in the medico-legal discourse of this time – which must insist upon the unique nature of the railway's power and effects in order to justify a specific pathology of the railway accident.

The shift in the latter decades of the nineteenth century from 'railway spine' to 'traumatic neurosis' is a crucial episode in the history of psychoanalysis and its theorisations of trauma and hysteria. Jean Martin Charcot, who exerted such a major influence on Freud's early writings, used Page's tentative suggestions that hysterical phenomena, including a form of 'psychical automatism', might result from railway injuries, to argue the case for 'male hysteria', which he opposed to 'neurasthenia'.[14] Hermann Oppenheim, writing in Germany, used the railway accident to theorise 'traumatic neurosis', emphasizing that symptoms may arise some time afer the event,[15] while Page noted that collapse from bodily injury, coincident with the injury itself, served to protect the accident victim from the 'prolonged and delayed manifestations of mental shock.'

Psychoanalysis could thus be said to draw its most fundamental concepts from the supposed effects of railway travel and railway accidents. Or, in slightly different logic and narrative, the railway becomes the symbol (rathar than the generator) of what Ian Hacking calls the 'psychologization of trauma' – the shift from *soma* to *psyche* which is central in Freudian psychoanalysis's account of itself.[16] The shift from *soma* to *psyche* suggests at the same time an equivalence between material and psychic life, as the railway accident comes to stand for traumas brought about by modernity and its new technologies, the forces of civilisation and industrialisation. In the early twentieth century and in Freud's work of the 1920s, in particular *Beyond the Pleasure Principle*, the theory of shock (which Benjamin was to use in his theories of modern structures of perception) re-emerges in the concept of shell-shock and war neurosis, the successors to 'railway shock'. As Ernst Bloch was to write: 'the modern war machine recalls the hellish face of the first locomotive.'[17]

Freud's biographers have also wanted to locate an intimate and indeed founding relationship between the railways and the founder of psychoanalysis. Marthe Robert speaks of Freud's profound travel anxiety which could be cured only by analysis.[18] Ernest Jones notes that Freud's train anxiety was cured by analysis but left its legacy in the form of a lasting anxiety about missing trains.[19] The violent anxiety experienced by Freud on train journeys is succeeded by something close to what I have been calling the logic of belatedness or missing: a slippage, so to speak, between subject and schedule.

Two train journeys taken by Freud and his family in his very early childhood become founding narratives of psychoanalysis. The first is the journey from Freiberg, the town of Freud's birth, to Leipzig, on which, passing through Breslau, he saw gas lamps burning and believed he was in Hell. His self-analysis, undertaken some 40 years later, revealed to him a fear of the loss of his mother, caught up with the actual loss of his birthplace. The second train journey was occasioned by the family's move from Leipzig to Vienna. Writing to Wilhelm Fliess in 1897, Freud stated that: 'between the ages of 2 and 2 and a half, my libido was stirred up towards *matrem*, namely on the occasion of a journey with her from Leipzig to Vienna, during which we must have spent the night together and I must have had an opportunity to see her *nudam*'.[20] This 'memory' founds the theory of the Oedipus complex and thus, in more than one way, the railway journey 'founds' psychoanalysis. Mothers, the anxiety of origin and the structures of deferral compete with machines and the violence of equivalence as principles of generation, of desire and of trauma.

III

Some 70 years before the publication of Walter Benjamin's essay on train-reading, Henry Mansel wrote a now well-known account of 'sensation fiction':

> The exigencies of railway travelling do not allow much time for examining the merits of a book before purchasing it; and keepers of

> bookstalls, as well as of refreshment-rooms, find an advantage in offering their customers something hot and strong, something that may catch the eye of the hurried passenger, and promise temporary excitement to relieve the dulness of a journey.[21]

'Railway fiction' emerged soon after the growth of the railway system in the early to mid-nineteenth century; cheap books or 'yellowbacks' were produced to be sold at the railway bookstalls which W.H. Smith began to take over in 1848. In the same year, Routledge initiated its 'Railway Library', and a number of other publishers swiftly entered the competition for the production of cheap editions. The growth of the railways is thus directly linked to the production of literature (popular and 'improving') with a mass circulation.

Although travel literature and other forms of non-fiction formed an important part of 'railway literature', a number of publishers had their greatest successes with modern fiction, including detective novels and the 'sensation novels' which are the object of H.L. Mansel's critique, quoted above. Mary Elizabeth Braddon's first novels, for example, appeared in yellowback editions in the 1860s and 1870s. In Richard Altick's account of nineteenth-century reading patterns: 'Every passenger train of the hundreds that roared down the rails in the course of a single day carried a cargo of readers, their eyes fixed on *Lady Audley's Secret* or *The Times*.'[22] Altick may or may not have intended a gendered division of reading here, but it is worth speculating in any case on the impact of the railways – which enabled women to travel far more easily and in far greater numbers than every before – on the 'feminisation' of literature.

Mansel's perception that the function of railway fiction, instanced by the sensation novel, is to provide 'temporary excitement to relieve the dulness of a journey' is repeated in discussions of travel throughout the century. His account of the sensation novel's 'preaching to the nerves' could be read as a displacement of the nervous shocks produced by railway travel to railway literature, in which case it comes close to Benjamin's assertion that the pleasure of the detective novel on the railway journey is that it provides a substitute anxiety to that produced by the journey itself.

For Mansel, books, bodies and machines are closely identified. Mansel links sensation novels with industrial processes: 'a commercial atmosphere floats around works of this class, redolent of the manufactory and the shop.'[23] They are condemned not only for their 'unnatural excitement' but for the 'rapidity of [their] production'. Such novels employ 'rapid and ephemeral methods of awakening the interest of their readers' and 'carry the whole nervous system by steam' (p. 487). They are limited to contemporary subjects because 'proximity... is one great element of sensation. It is necessary to be near a mine to be blown up by its explosion; and a tale which aims at electrifying the nerves of the reader is never thoroughly effective unless the scene be laid in our own days and among the people we are in the habit of meeting' (pp. 488–9). Novels, in this model, sound remarkably like trains, with the difference that in the process of reading 'nervous shock' can become a pleasurable sensation. And although Mansel represents the train journey as dull, the train-book equivalence he constructs suggests that the railway nonetheless continues to be imbued with a high potential for danger.

The railway accident played a significant part in sensation fiction. It is central, for example, to Mrs Henry Wood's *East Lynne* (1861) and Mary Braddon's *John Marchmont's Legacy* (1863) as a device allowing for the exploration, and destabilising, of identities and of memory and its loss. Wood's plotting uses the railway accident to cut the novel in two, making the second half a punitive repetition of the first for the heroine. At the centre of the novel is a railway accident that kills the errant Lady Isabel Vane's illegitimate child and radically disfigures Lady Isabel herself. The rescuers at the scene of the railway accident had left her for dead, but

> the surgeons, on further inspection, had found life still lingering in her shattered frame.... She remained three months in the hospital before she could be removed. The change that had passed over her in those three months was little less than death itself: no one could have recognized in the pale, thin, shattered, crippled invalid, she who had been known as Lady Isabel Vane.[24]

The railway accident 'shatters' and 'disfigures' the human form and face – dissolving identities in an instant. Presumed dead in the accident, Lady Isabel is able to return, unrecognized, to her former home: to act as governess to the children she had left behind; to witness, in agonies of remorse and jealousy, the happiness of her former husband and his new wife; and to die of consumption and grief.

In *John Marchmont's Legacy*, brain fever and amnesia follow a skull fracture sustained by the hero in a railway accident:

> For eleven weeks after that terrible concussion upon the South-Western Railway, Edward Arundel had lain in a state of coma, – helpless, mindless; all the story of his life blotted away, and his

> brain transformed into as blank a page as if he had been an infant lying on his mother's knees.[25]

Mary Braddon's novel, in line with a number of theories of hysteria, represents female pathologies as innate, sexually charged and perhaps hereditary, while male pathologies are the product of external traumatic forces. Yet the distinction between endogenous and exogenous aetiologies is also disturbed by the equivalence between machine and body, and the complex question of identification in the discourse of 'sensation' (a relay between nervous systems, as in Mrs Oliphant's famous pronouncement that in the sensation novel 'the reader's nerves are affected like the hero's').[26] At the same time, the gender of psychopathology is troubled through the discourses of hysteria, which arise in large part out of the study of railway trauma.

If the sensation novel drew, literally or figuratively, on the railway accident and the traumas of train travel, what did the railways offer to writers of detective fiction, its generic successor? The standardisation of time, which directly resulted from railway travel, is clearly central to a genre dependent on establishing alibis and times of death. The railway timetable becomes the detective and the detective novelist's guide – the text of standardised time. Train travel also comes to represent the combination of rationality and the unknown which corresponds to the play between ratiocination and intuition in the classical detective narrative.

Furthermore, the railway created not only the possibility of combining an enclosed space with mobility and changing views but new spaces of darkness and isolation: the tunnel (perhaps the bourgeoisie's only experience of a subterranean world familiar to a number of industrial workers) and the compartment without corridors. Railway detective fiction charts the changing architecture of trains and often based itself on 'true railway crimes', such as the murder of one Mr Gold in 1881, a businessman, shot as the train passed through one tunnel and thrown out of the train in the next. *The Times* reporter concluded that 'the ordinary railway compartment into which English men and women pen themselves off by twos or threes affords facilities to the garotter or the murderer of which we may be thankful that advantage has been so seldom taken.'[27] *Reynolds's Newspaper* also linked this murder with that of a Mr Briggs in 1864 and noted the particular dangers of the closed compartment: 'The man who sits opposite may be a madman or an assassin. The train thunders along, and the roar, deadening all other sounds, conceals a shriek as effectually as the deepest dungeon of the Bastille.'[28]

I return now to my three logics – equivalence, belatedness, separation/inversion – and to their working through in crime and detective fiction. Any discussion of railways and novels must include Emile Zola's *La Bête Humaine* (1889/1890), where murder is modernised and made inseparable from machine culture, and where accident and psychopathology become indivisible. Roubaud, the deputy stationmaster at Le Havre, discovers that his young wife Séverine has been having a long-standing affair with her godfather President Grandmorin, a former judge who is now on the board of directors of the railway company for which Roubaud works. He murders the President on the train on which they are both travelling, a murder which is glimpsed by Jacques Lantier, a train driver tormented by his pathological desire to kill women. Jacques sees the murder committed as he stands at a spot near a railway tunnel:

> The whole line of carriages followed, the little square windows, blindingly light, making a procession of crowded compartments tear by at such a dizzying speed that the eye was not sure whether it had really caught the fleeting visions. And in that precise quarter-second Jacques quite distinctly saw through the window of a brilliantly lit coupe a man holding another man down on the seat and plunging a knife into his throat, while a dark mass, probably a third person, possibly some luggage that had fallen from above, weighed down hard on the kicking legs of the man being murdered.[29]

Finding the body of the President thrown from the carriage, Jacques's 'itch for murder intensified like a physical lust at the sight of this pathetic corpse' (p. 73). The tableau he has seen does not fall into place, however, until later in the novel, when he and Séverine have become lovers. She describes the murder to him when they are in bed, revealing that she was the dark mass holding down the President; her body was the medium through which the shock of the stabbing was felt. 'I didn't see anything,' she tells Jacques, 'but felt it all, the impact of the knife in his neck, the long tremor of the body, and death in three hiccuping gasps, like the spring of a clock breaking.... Oh, I can still feel a sort of echo of his death-struggle in my own limbs' (pp. 231–2). It is not the sight of murder but its physiological sensations – its shock effects – that excite Jacques so intensely; after her confession 'they possessed each other... with the same agonising pleasure as beasts disembowelling each other in mating' (pp. 234–5).

Throughout the novel Zola represents the connection between mechanical agitation and sexual arousal which Freud was later to theorise for psychoanalysis, as well as the ways in which the murderer represents his own death through the death of another. The train itself becomes the focal point of the interchange between the ostensibly disparate systems of machines and bodies: the collapse of the distinction also collapses the distinction between sexual desire and the desire to destroy. Jacques finally murders Séverine, 'the same kind of stab as for President Grandmorin, in the same place, with the same savagery. At that moment the Paris express dashed past, so noisy and fast that the floor shook, and she was dead as if this storm had struck her down' (p. 331). The repetition of the act – 'the same kind of stab ... in the same place, with the same savagery' – is a reinscription on the body and in the writing of the novel itself. Repetitive murder, of the kind that obsesses Jacques, is, as Mark Seltzer notes, defined in the conjunction of compulsive sexualised violence on the one hand, with technological system, schedule and routine (such as the railway is governed by) on the other.[30]

Zola's narrative is itself obsessively and repetitively punctuated by railway murders and disasters. In addition to the murder on the train, the novel describes a perilous journey through a snow-storm, in the aftermath of which Flore, also in love with Jacques, resolves to kill Jacques and Séverine. She effects a railway disaster by drawing a horse-drawn cart in front of their train; she then commits suicide in a railway tunnel. At the close of the novel, Jacques and his assistant fight to the death under the train they are driving. The runaway train speeds towards the future and destruction with a cargo of soldiers bound for the Franco-Prussian war. Their bodies will presumably be broken and fragmented by one of two technologies of modernity: the war machine or the railway.

My next example, in which I locate what I have called the logic of belatedness, is a reworking of Zola's novel in which equivalence and simultaneity are rejected in favour of various forms of 'missing'. Georges Simenon's *L'homme qui regardait passer les trains* (*The Man Who Watched the Trains Go By*) is the story of Kees Popinga, a middle-aged Dutch bourgeois who, on discovering that his employer's business has collapsed and that his career has been founded on sand, suddenly abandons family life, takes a train to Amsterdam, murders his employer's mistress when she refuses his advances and escapes to Paris. Simenon uses the image of passing trains to suggest the sexual violence always potential in Popinga and his murderous innocence:

> And if, in introspective mood, he had set himself to discover if there were a streak of wildness latent in his mental make-up, nothing is less likely than that he would have thought of a certain queer,

> half-guilty feeling that crept over him whenever he saw a train go by – especially a night-train, with all its blinds down, rife with mystery.[31]

The most evocative image of the thriller genre – the night-train, 'with all its blinds down' – thus becomes, in Simenon's crime novel, the figure of (male sexual) fantasy and unacknowledged desire itself: the 'dimmed lights and close-drawn blinds' of night-expresses are linked for Popinga with the shadows on the blinds on Groningen's one 'gay house'. On the run after committing murder, Popinga hides out next to a marshalling yard. The movement of the trains at night retroactively produces desire for a prostitute with whom he has recently spent a chaste night:

> before getting into bed, [he] gazed for several minutes at the glimmering expanse of railway lines, red and green lights, a long, black goods-train rumbling by. His thoughts kept harking back to Jeanne, and now he found a curious pleasure in picturing amorous contacts which, when they were available, he had foregone. (p. 82)

Popinga's former conventionality is imaged through the mores of railway travel: 'On railway journeys I had to pretend to read a book or look out of the window, instead of showing any interest in those around me, and to wear gloves, uncomfortable as they might be – because it's the right thing to do when travelling.' (p. 129) Trains, that is, arouse desire: the conventions of travelling demand its repression and its sublimation through reading or looking through the glass at the world outside. A similar tension exists between the train's transgressive properties – its movements through the night, 'the rhythmic panting of engines' (p. 86), and its crossing of borders – and the routinised spatio-temporal world of railway travel, mirrored in the obsessive calculations and grids – maps, chess-games, accounting – which preoccupy Popinga, managing clerk turned murderer. At the close of the novel, Popinga attempts suicide by lying across a railway-track. He is caught and arrested, 'clad only in his overcoat': 'In the course of the night he was roused and made to put on a ticket-collector's uniform which was much too small for him.' (p. 186)

Simenon's reworking of Zola's novel splits the machine–body complex. Popinga's desire, aroused by the sight and sound of the trains in the shunting-yard, is too late, a connection missed. Popinga even 'misses' his own death, lying across the wrong line:

> With a rattle and a roar the light bore down on him, and the roar swelled to such a din as he had never heard in his life before, so tremendous that he almost fancied he was dead already.

> But presently there came a sound of voices, followed by an eerie silence. Only then he realized that the train had stopped, on the other line, and saw two men climbing down from the engine, and carriage windows being let down. (p. 185)

By placing Popinga on the wrong track, the narrative prevents body and machine from meeting and constructs an alternative nemesis, with Popinga incarcerated in an asylum.

In coming to our third 'logic', the imperative of separation/inversion, we encounter, in another term of de Certeau's, the 'spatialising frenzy' of the detective novel.[32] The space of the tunnel, as we have seen, is the most consistent nexus of narrative activity and anxiety. Subterranean and viewless, it is frequently figured as the site of lawlessness and of action without witness. In the history of railway construction, tunnels accounted for the largest number of fatalities and injuries among workers. They were also the most frequent sites of railway 'accidents', thus threatening not just the working class but the travelling middle classes. The spaces of anxiety from the start, it would seem as if fear of injury or death from the railway accident commuted itself to, or began to run concurrently with, fear of injury or death, or loss to property or propriety, at the hands of one's *compagnons de route*. Certainly, it is the latter or later anxiety which motivates the narratives of railway detective fiction.

The journey through the tunnel precludes the view from the outside in and the inside out. The spatial structures of the closed compartment and the train without corridors create mysteries dependent less on sight and sightlessness than on the seemingly inexplicable movement of people or objects into and out of an enclosed space which moves through space and time at an unprecedented rate. The virtual interchangeability of bodies, living and dead, and luggage (particularly boxes of various kinds – jewel boxes, despatch boxes) in railway detective fiction plots suggests something of the felt commodification of persons on the railway, reminding us of Ruskin's complaint that the railway journey turned the traveller into a parcel, ticketed and despatched from one place to another.[33]

Sherlock Holmes is, as Benjamin suggests, closely identified with the railway. Yet, and despite the fact that at least two thirds of the Sherlock Holmes stories contain train journeys and that Holmes is a figure shot through with nervous sensations, Conan Doyle rarely makes the railway the site of crime. The railway journey tends to be used, as Benjamin suggests, to mark narrative duration and movement, or as a narrative space in which Holmes can impart to Watson the contexts of crimes on outward journeys and their solutions on return. Holmes's famous comment that it is the country rather than the city in which criminality flourishes is made on a train journey, the view out of the window allowing for a kind of inversion – in this case of the traditional city/country divide.[34]

For one among numerous examples of 'closed carriage mysteries' we could turn to Conan Doyle's 'The Story of the Man with the Watches' (not a Holmes story).[35] Here the closed compartment mystery substitutes one form of 'crossing' for another, hypothesising a movement between parallel trains, the traversal of separate spaces, only in order to reveal that the 'crossing' was in fact sexual, the murdered man having entered the carriage as a woman. The movement from one train to another could not have occurred because the two trains do not in fact run in parallel or at the same speed; the explanation is thus not that of the straight lines of the railway but the confusion of sexual 'deviance' or 'inversion'.

Doyle, in this story, explicitly turns his back on the view into the train carriage from a train passing alongside. Such a view is, however, central to railway detective fiction, most famously, perhaps, in Agatha Christie's *4.50 from Paddington*. The fleeting landscape is transformed by the violent act into a tableau, perceptible only as a result of the pure contingency of being at a particular point at a specific time, as Mrs McGillicuddy, travelling on a train on her way to stay with her friend Jane Marple, sees a woman murdered in a compartment of a train running parallel to her own:

> At the moment when the two trains gave the illusion of being stationary, a blind in one of the carriages flew up with a snap. Mrs McGillicuddy looked into the lighted first-class carriage that was only a few feet away.
>
> Then she drew her breath in with a gasp and half-rose to her feet.
>
> Standing with his back to the window and to her was a man. His hands were round the throat of a woman who faced him, and he was slowly, remorselessly strangling her. Her eyes were starting from their sockets, her face was purple and congested. As Mrs McGillicuddy watched, fascinated, the end came; the body went limp and crumpled in the man's hands.
>
> At the same moment, Mrs McGillicuddy's train slowed down again and the other began to gain speed. It passed forward and a moment or two later it had vanished from sight.

> Almost automatically Mrs McGillicuddy's hand went up to the communication cord, then paused, irresolute. After all, what use would it be ringing the cord of the train in which *she* was travelling? The horror of what she had seen at such close quarters, and the unusual circumstances, made her feel paralysed. *Some* immediate action was necessary – but what?[36]

The transformation of the 'flying countryside', as seen from the train window, into the interior space, as seen from the parallel railway compartment, inaugurates the plot: a bourgeois family drama situated in a country house which is encircled by the railway. The intimacy of Mrs McGillicuddy's view from one compartment window into another (what allows her, in de Certeau's terms, 'to see') rests on 'the illusion of being stationary', in the same way that she is drawn to pull the communication cord by the momentary illusion that the proximity of the two trains in space and time is equivalent to an identity of their systems. The logic of separation here runs alongside, or perhaps gives way to, a logic of equivalence, as Miss Marple brings about a resolution of the mystery (a mystery which is initially that of a murder without a body): first by repeating the course of the train journey herself and finally by forcing the murderer into a repetition of his murderous gestures:

> [Miss Marple] gave a sudden gasp and began to choke. 'A fish bone', she gasped out, 'in my throat.' ... Mrs McGillicuddy gave a sudden gasp as her eyes fell on the tableau in front of her, Miss Marple leaning back and the doctor holding her throat and tilting up her head.
>
> 'But that's *him*', cried Mrs McGillicuddy. 'That's the man in the train ...' (p. 215)

Ethel Lina White's *The Lady Vanishes* (better known through Hitchcock's film version) is a train mystery which largely excludes the outside, a Balkan landscape, making its substitutions within and between train compartments.[37] Thus Iris Carr is alerted to the disappearance of Miss Froy, the English spinster who has befriended her, when her seat becomes and remains empty; she is confirmed in her belief in a conspiracy when Miss Froy's place in the compartment is taken by a woman wearing her clothes. Miss Froy is, in fact, lying heavily bandaged and under guard in the next door compartment, disguised by her abductors as a maimed accident victim.

Iris, herself the recent recipient of a blow to the head, cannot be entirely sure of what she knows and sees; the reality of Miss Froy's existence is secured for Iris by the sight of her name written on a pane of glass. In White's novel, the glass separates the compartment from the corridor: Iris and Miss Froy occupy corner rather than window seats. Hitchcock, in his 1938 film *The Lady Vanishes*, transferred the signature from the corridor glass to the train window, adding for good measure a label from the packet of tea ('Harriman's Herbal Tea') preferred by Miss Froy. The train window – the windowpane which connects, separates and inverts inside and outside – becomes the site on which identity is both secured and effaced, as signature and label respectively are revealed and vanish, adhere and are whirled away. With a knowing gesture to Freudian psychology, the film-script plays with the concept of 'substitution' as both an unconscious mental act and as a villainous ploy: 'We believe there's been a substitution, doctor', Iris and her one ally on the train declare. There may even be an allusion to the name of Freud himself: 'Froy – it rhymes with joy', proclaims the young-old English spinster, as she spells out her curious name on the windowpane of the restaurant car.

My final example of the logic of separation/inversion is Patricia Highsmith's *Strangers on a Train* (1950). Here one train traveller finds not only that he has swapped stories with a fellow traveller, but that he is expected to swap murders: 'We murder for each other, see? I kill your wife and you kill my father! We meet on the train, see, and nobody knows we know each other! Perfect alibis!'[38] The psychopathic Bruno kills Guy's estranged wife and then demands of Guy that he kill Bruno's hated father. The structure is 'criss-cross', as Bruno reiterates in Hitchcock's film version of the novel, or, in the terms of de Certeau's account of railway travel, 'chiasmic'. Here the structures of exchange become caught up in a logic as murderous as that of the equivalence and compulsion to repeat which drive the characters in Zola's novel.

IV

In the 1930s, Walter Benjamin's account of railway anxiety is anachronistic – and deliberately so. Representing the anxieties of the journey in 'mythic' form, he condenses the cultural history of a century in its movement from wonder and fear of the new technology of rail travel to its 'naturalisation', embodied in the traveller who reads during the journey. But Benjamin leaves open, or opens up, the space of anxiety; a subliminal anxiety, which he terms 'mythic', specific to the technologies of travel and always capable of being reawakened. It is appropriate, then, that on the journey the traveller should turn to the 'mythic' literature of nineteenth-century detective writing, a genre near-contemporaneous with the expansion of the railways and railway travel.

And yet Benjamin's account is also fitted to his times. It is in fact the cinema that reinvents both the railway journey and detective narrative for the twentieth century. It perhaps reinvents railway trauma itself, as we recall the 1895 film *Arrival of a Train*, in which the train is said to have appeared to its first spectators to be breaking through the screen to run them down; many are reported to have cried out in fear or to have fainted. The story may be exaggerated or even apocryphal, but the striking point is that it becomes the founding myth of cinema, bringing together film, train and trauma.[39] As a number of critics have noted, the experience of the cinema seemed in many ways to be an extension of the experience of rail travel, with its panoramic view from the train window, the traveller, as Benjamin Gastineau wrote in 1861, 'gazing through the compartment window at successive scenes.'[40] Film brings the detective novel back into the visual field – much of Hitchcock's cinema is unthinkable without the train. In the 1930s, of course, new anxieties emerged as trains crossed Europe's increasingly contested and fraught borders, but these are stories which I cannot explore further here.

Notes

1 Walter Benjamin, 'Kriminalromane, auf Reisen', in *Gesammelte Schriften*, IV.1 (Frankfurt am Main: Suhrkamp, 1989), pp. 381–2.

2 Benjamin's outline for a detective novel is published in *Schriften*, VII. 2, 845–51. See also Ernst Bloch, 'A Philosophical View of the Detective Novel' (1965), in Ernst Bloch, *The Utopian Function of Art and Literature: Selected Essays*, trans. Jack Zipes and F. Mecklenburg (Cambridge, Mass: MIT Press, 1988), pp. 245–64; and Siegfried Kracauer, *Der Detektiv-Roman: Ein philosophicher Traktat*, in *Schriften*, 1 (Frankfurt am Main: Suhrkamp, 1971), pp. 103–204. For further discussion of Frankfurt School theorists and their interest in detective fiction, see David Frisby, 'Walter Benjamin and Detection', *German Politics and Society*, 32 (Summer 1994), 89–106, and 'Between the Spheres: Siegfried Kracauer and the Detective Novel', *Theory, Culture and Society*, 9.2 (1992), 1–22.

3 Benjamin, 'Kriminalromane, auf Reisen', p. 381. Translations from Benjamin are my own.

4 Mark Seltzer, *Bodies and Machines* (New York and London: Routledge, 1992), and *Serial Killers* (New York and London: Routledge, 1998).

5 Wolfgang Schivelbusch, *The Railway Journey: Trains and Travel in the Nineteenth Century*, trans. Anselm Hollo (Oxford: Basil Blackwell, 1980), p. 44.
6 Michel de Certeau, 'Railway Navigation and Incarceration', in *The Practice of Everyday Life* (Berkeley, CA: University of California Press, 1984), p. 112.
7 Benjamin, 'Kriminalromane, auf Reisen', p. 381.
8 Sigmund Freud, 'Three Essays on the Theory of Sexuality' (1905), *Standard Edition*, VII (London: The Hogarth Press and the Institute of Psycho-Analysis, 1953), pp. 201–2.
9 *The Book of Health*, ed. Malcolm Morris (1884). Quoted by Schivelbusch, *The Railway Journey*, p. 118.
10 See for example the accounts of the nervous system in George M. Beard's *American Nervousness: Its Causes and Consequences* (New York: Putnam, 1881).
11 Max Nordau, *Entartung* 2 vols (Berlin: Duncker, 1892), translated as *Degeneration* (London: William Heinemann, 1895), p. 41.
12 John Eric Erichsen, *On Railway and Other Injuries of the Nervous System* (London: Walton and Maberley, 1866), p. 2.
13 Herbert W. Page, *Railway Injuries: with Specific Reference to those of the Back and Nervous System in their Medico-Legal and Clinical Aspects* (London: Charles Griffin, 1891), p. 28.
14 See J.-M. Charcot, *Leçons sur les maladies du système nerveux* (Paris: Progrès Medicale, 1886–7); and Page, *Railway Injuries*, p. 50.
15 See Hermann Oppenheim, *Diseases of the Nervous System: A Text-Book for Students and Practitioners of Medicine*, trans. Edward E. Mayer (London: Lipincott, 1904), especially pp. 778–94.
16 Ian Hacking, *Rewriting the Soul: Multiple Personality and the Sciences of Memory* (Princeton University Press, 1995), especially pp. 183–97. See also E. Fischer-Homberger, *Die traumatische Neurose: Vom somatischen zum sozialen Leiden* (Berne, Stuttgart, Vienna: Hans Huber, 1975).
17 Ernst Bloch, *Spuren* (Frankfurt am Main: Suhrkamp, 1969), p. 161.
18 Marthe Robert, *The Psychoanalytic Revolution: Sigmund Freud's Life and Achievement*, trans. Kenneth Morgan (London: George Allen and Unwin, 1966), especially pp. 111 and 144–7.
19 Ernest Jones, *The Life and Work of Sigmund Freud*, ed. and abridged by Lionel Trilling and Steven Marcus (Harmondsworth: Penguin, 1974), pp. 41 and 264. See also George Frederick Drinka, *The Birth of Neurosis: Myth, Malady and the Victorians* (New York: Simon and Schuster, 1984), especially pp. 198–222.
20 *The Complete Letters of Sigmund Freud to Wilhelm Fliess 1887–1904*, ed. J.M. Masson (Cambridge, Mass: Harvard University Press, 1985), p. 268 (letter of 3 October 1897).
21 H.L. Mansel, 'Sensation Fiction', *Quarterly Review*, 113 (1863), 485.
22 Richard Altick, *The English Common Reader* (Chicago: University of Chicago Press, 1957), p. 89.
23 Mansel, 'Sensation Fiction', p. 483.
24 Mrs Henry Wood, *East Lynne* (London: Dent, 1994), p. 327.
25 Mary Elizabeth Braddon, *John Marchmont's Legacy* (London: Simpkin, Marshall, Hamilton, Kent, 1892) p. 125. For an interesting discussion of the novel in relation to Victorian psychology and psychiatry, see Sally

Shuttleworth, 'Preaching to the Nerves', in *A Question of Identity: Women, Science and Literature,* ed. Marina Benjamin (New Brunswick, NJ: Rutgers University Press, 1993), especially pp. 214–20.

26 Margaret Oliphant, 'Sensation Novels', *Blackwood's,* 91 (May 1862), 572.
27 *The Times,* 7 July 1881, p. 9.
28 *Reynolds's Newspaper,* 3 July 1881.
29 Emile Zola, *La Bête Humaine,* trans. Leonard Tancock (Harmondsworth: Penguin, 1977), p. 70.
30 Mark Seltzer, 'Serial Killers (II): The Pathological Public Sphere', *Critical Inquiry,* 22 (Autumn 1995), 122.
31 Georges Simenon, *The Man Who Watched the Trains Go By,* trans. Stuart Gilbert (London: Pan, 1948), p. 5.
32 Michel de Certeau, 'Spatial Stories', in *The Practice of Everyday Life,* p. 118.
33 Quoted by Schivelbusch, *The Railway Journey,* p. 45.
34 Arthur Conan Doyle, 'The Adventure of the Copper Beeches', in *The Penguin Complete Sherlock Holmes* (Harmondsworth: Penguin, 1981), pp. 322–3.
35 Arthur Conan Doyle, 'The Story of the Man with the Watches', *Strand* (August 1898).
36 Agatha Christie, *4.50 from Paddington* (London: Pan, 1974), pp. 7–8.
37 Ethel Lina White, *The Lady Vanishes* (London: Dent, 1987) originally published as *The Wheel Spins* (London: Collins, 1936).
38 Patricia Highsmith, *Strangers on a Train* (Harmondsworth: Penguin, 1974), p. 30.
39 For further discussion of trains and film, see Ian Christie, *The Last Machine: Early Cinema and the Birth of the Modern World* (London: British Film Institute, 1994) and Lynne Kirby, *Parallel Tracks: The Railroad and Silent Cinema* (University of Exeter Press, 1997).
40 Benjamin Gastineau, *La vie en chemin de fer* (1861), quoted by Schivelbusch, *The Railway Journey,* p. 63.

16
Open Letter to Detectives and Psychoanalysts: Analysis and Reading

Patrick ffrench

There is an evident structural parallel between the activity of the psychoanalyst and that of the detective. Both *analyse* a texture of manifest clues or symptoms in order to find the hidden or latent truth that lies behind this surface.[1] However, this structural parallel, and the model of reading which underlies it, are called into question if we pay attention to the letter of the texts of Freud and Lacan. For if Freud proposes a theory of reading, this theory undermines the notion of uncovering the hidden truth, of revealing the hidden crime; what emerges from the account of the reading or interpretation of the dream is the pre-eminence of *syntax* over *content*. Deconstructive critics such as Jacques Derrida or Jeffrey Mehlman have argued that the Freudian recognition of the pre-eminence of syntax over content is invariably betrayed by an eventual postulation of truth, presence, a *semantic* principle, or drive to locate meaning (in one place). From this perspective, the enjoyment of the fold, of the syntax of detective fiction, is ultimately let down by the revelation of the crime or the criminal. Truth is disappointment. Moreover, from a deconstructive perspective, reading is not the passive reception or consumption of a pre-existing meaning, but the production of meaning through a transferential play with the text and with other texts. The notion of reading must also involve an account of transference and this also complicates the parallel proposed above, based on a simple model of reading as uncovering. If psychoanalysis, detective fiction and deconstruction propose an enjoyment (*jouissance*) of the fold, of syntax, they also inevitably end up posing a semantic core, a truth, the *meaning* of their discourses.

If we read psychoanalysis *with* detective fiction, at its inaugural moment, we will see that right at the beginning detective fiction proposes an insight which also undermines the model of reading on which the notion of the revelation of monovalent truth is based. What will be proposed here is that there is a different parallel behind the *evident* one proposed above, and it is premised on a theory of reading. The question is: does Freud offer a model or a theory of reading? This seemingly theoretical question is also one which addresses contemporary culture. For if contemporary society is obsessed with the myth and ideology of the detective (one would have to execute a Foucauldian analysis of precisely when in modern history the detective appeared as an idea), it is as readers of texts and of films that we engage in this obsession. Does not the novel, through identification with a hero or heroine, invite the reader into the position of the detective, and does not film, through the effect described as *suture*,[2] stitch the reader into the texture of the film, precisely as reader of visual and auditory clues, symptoms of the trauma past or to come? To look at how psychoanalysis might inform a reading of detective fiction, it is first necessary, then, to ask if psychoanalysis proposes a theory of reading.

The solidity of the parallel proposed above as evident would be dissolved at a stroke if one were to propose that psychoanalysis were in some sense founded on or inaugurated by a writer of detective fiction. Through an effect which Freud might describe using the term *Nachträglichkeit* – 'deferred action' – the psychoanalytic project is haunted by its previous configuration in detective fiction. It finds itself[3] curiously defined but also challenged in this literary form whose inauguration takes place 60 years or so before its time, in 1841:

> The mental features discoursed of as the analytical, are, in themselves, but little susceptible of analysis. We appreciate them only in their effects. We know of them, among other things, that they are always to their possessor, when inordinately possessed, a source of the liveliest enjoyment. As the strong man exults in his physical ability, delighting in such exercises as call his muscles into action, so glories the analyst in that moral activity which *disentangles*. He derives pleasure from even the most trivial occupations bringing his talents into play. He is fond of enigmas, of conundrums, of hieroglyphics; exhibiting in his solutions of each a degree of acumen which appears to the ordinary apprehension preternatural. His results, brought about by the very soul and essence of method, have, in truth, the whole air of intuition. The faculty of re-solution is possibly much invigorated by mathematical study, and especially by that highest branch of it, which, unjustly, and merely on account

> of its retrograde operations, has been called, as if *par excellence*, analysis. Yet to calculate is not in itself to analyse.[4]

The writer is Edgar Allan Poe, at the outset of 'The Murders in the Rue Morgue', which has been described as inaugural of the detective story. It is the first of a trilogy of stories featuring the detective Auguste Dupin, of which the third is 'The Purloined Letter', studied by the psychoanalyst Marie Bonaparte in France[5] and after her by Jacques Lacan,[6] who inaugurates his *Ecrits* with it, before being taken to task upon his analysis by Jacques Derrida,[7] who thus inaugurates an industry of Purloined Letter theorizations and ripostes.[8] What is at stake in this series of displaced inaugurations is a theory of reading.

Poe continues, in this text, to demonstrate, via an analogy with the game of draughts, the necessarily superior quality of an analysis which bases itself upon an assessment of the intelligence of the adversary, on an identification with this intelligence: 'The analyst throws himself into the spirit of his opponent, identifies himself therewith, and not infrequently sees thus, at a glance, the sole methods (sometimes indeed absurdly simple ones) by which he may seduce into error or hurry into miscalculation.'[9] The detective, Dupin in this case, becomes a pretextual model for the psychoanalyst attempting to outwit the repressive forces of the patient's ego and seduce or hurry the patient into a symptomatic slip. Or, in the later form of the detective novel, under the reverse influence of psychoanalysis, the detective becomes both analyst and patient, seducing the criminal via a form of transference, into confession or capture, as a ruse to discover the truth of the past, their own, or that of their love-object.

But if the parallels are sound and the lesson clearly learnt from detective story to psychoanalysis and back to detective novel, between Dupin's identificatory ruses, Freud's transferential sparring with his patients, and the modern detective's search for self, there are fundamental differences regarding the modes of reading, particularly modes of reading material objects, and over this aspect it seems possible to suggest a critique of both contemporary cultural forms and of psychoanalysis, by way of a reading of Poe.

Both Poe and Conan Doyle's detectives work through what might be called a Bloomian, agonistic intertextuality with the agents of the law, or, more simply, through dialectics: critique and transgression of the methods of the police. Poe, like his admirer Mallarmé, is an adept intertextual reader of the popular press, and, in 'The Mystery of Marie Roget', solves the crime more or less through a critical reading of the newspapers.[10] But not all detectives and not all readers are equipped with the necessary *acumen* to play this game of interpretation and counter-interpretation. The ordinary detective contents him or herself with being a reader of signs, like the psychoanalyst. In detection, however, semiology only goes so far. Julia Kristeva points out in one of her earliest essays, a review of Barthes's *The Fashion System*, that semiology as a science eventually reaches tautological closure with the recognition that it is itself a result of the ideology of the sign which it studies.[11] This closure, or circle, defines the limits of the detective who does not supplement his reading of material clues with a transferential dynamic. In the history of semiology and its developments, Kristeva and Barthes move towards psychoanalysis.[12] The dull detective remains stumped before an unbreakable code and psychoanalysis does not work without transference. Freud stipulates that the two principal discoveries of psychoanalysis are interpretation and transference. The second enables the success of the former which on its own is blocked by repression.

This might be mapped onto a very general sketch of the genealogy of the detective fiction form after its majestic and, it seems to me, largely unassimilated inauguration with Poe. Conan Doyle's Sherlock Holmes represents an intensification of the aspect of the expert reading of material signs, the seemingly 'preternatural' use of logic, at the expense of the transferential, identificatory aspect, but not completely. This latter aspect, less prominent in the Holmes stories, becomes intensified in a writer like Chandler, with the major and important distinction that the subject whose truth the detective is in search of is the detective himself. Psychoanalysis seems to have become a hidden or acknowledged subtext for the detective, who is now not so much in search of the truth of the crime he or she is investigating, as in search of the truth or trauma of a past, often his or her own past. The detective, like the analyst, must engage in transference with the criminal or the patient, must identify with the criminal in order to trace the path back to the original trauma. To take another example, the problem for Patricia Highsmith's psychotic hero Ripley is how to negotiate a series of shifts in identity and avoid a confrontation with himself. In this case, the detective, as narrator, becomes the criminal: this is transference gone wild to the extent of the destruction of identity and the annihilation of the middle man, a transgression of the rules of the genre. In more restrained cases, the criminal becomes the double of the detective. Thus Decker in *Blade Runner*, part 1950s detective fiction, part sci-fi fantasy, tracks down and kills the 'replicants' in a search for his own past, his own non-existent memories. The *dénouement* suggests that the last victim will be himself, for he too is a replicant; his trauma is the lack of origin, the displacement of the dreams he thinks are his own from their actual source. In this later manifestation, the detective becomes both analyst and patient, the criminal is revealed as the double on the way to self-recognition, a circularity which travels by way both of the reading of material clues and transference, identification, with the criminal other. Both analyst and detective, in this case, are after a cure.

If a certain epoch which might be loosely identified as modern is defined as that of the sign and its hero is the detective, then contemporary postmodern culture occupies an ambivalent position between a culture of signs which stand for ideas, and a culture of visual or auditory surfaces and shocks which mean nothing but their own violence. The contemporary detective becomes more and more a tragic victim of the loss of meaning, the lack of crime in a society in which violence is all. The replacement of *detective* fiction by *crime* fiction as the major obsession of contemporary postmodern culture, from *Bonnie and Clyde* and *Badlands*, to *Reservoir Dogs* and *Natural Born Killers*, and so on ad nauseam, a phenomenon which might be for simplicity's sake be called the 'Tarantino effect', is determined by a shift from the sign to violence or affect, foreclosed from the previous era and back with a vengeance.

Psychoanalysis informs semiology as it informs the philosophy of the detective. For Freud, the objects which feature in dreams, or phantasies, are signifiers whose true meaning lies in their standing for, their representation of, something other. But Freud's theory of reading is complex and involuted and provides, with certain limitations, a manual for the detective as semiologist.

In *The Interpretation of Dreams* Freud sets out the two traditional models for the reading of dreams: the symbolic method and the decoding method.[13] While the symbolic method simply transposes the dream as a whole, replacing it with a meaningful content, decoding translates each sign in the dream into another sign, via a kind of cryptography. Both of these modes of reading – transposition of the whole story and translation of individual elements – are, apparently, misguided. But Freud does appeal to them, against 'the philosophers and the psychiatrists',[14] taking elements from each, but favouring the latter approach of decoding, and 'regarding dreams from the very first as being of a composite character, as being conglomerates of psychical formations.'[15] The key difference with Freud's new science, it seems, is that the task of interpretation is imposed upon the dreamer him- or herself, via the impartiality of the analyst. Thus a theory of reading is always doubled, and psychoanalysis owes its difference to this, by a theory relating to identification and transference. As Poe proposes, calculation is not enough; we need the superior, identificatory play of a Dupin to accompany the dull sign-transposition of an analyst like the policeman G. in these tales, or Lestrade in Conan Doyle. We need, in other words, the subtle transferential dynamics of Barthes as writer to accompany the Barthes of semiology, if we are going to get anywhere, in draughts or in detection.

To return to reading, Freudian theory is also complex in proposing the involuted figures of condensation and displacement in its approach to the signifier, or object, and this suggestion of the plurality and mobility of signs, which derives from the Freudian theory of reading, informs structuralist and post-structuralist readings of texts and culture and, prefigured in Poe, it should, at least, inform the practice of the detective. However, the affirmation of a plural and mobile approach to reading seems belied by the quite linear and dogmatic form of what Freud proposes as truth. As Derrida remarks in 'Le facteur de la vérité', everything that does not belong to the primary core of the Oedipus complex belongs to 'secondary revision', to distortion or veiling.[16] What Freud gives with one hand he takes away with another, if we are looking for a theory of reading. On the one hand Freud offers a complex, plural and mobile perspective on the function of the signifier, while on the other he performs a simple transposition-type substitution of one content for another: *this* dream means *that*. According to Derrida, Lacan performs a similar operation in his commentary on 'The Purloined Letter' which as he (Lacan) presents it *means* the 'pre-eminence of the signifier (the letter) over the subject.'[17] In a key article, Jeffrey Mehlman proposes that precisely what is repressed by Freud in his specimen dream of Irma's injection in *The Interpretation of Dreams* is the original and radical notion of the dream as a syntax, replacing the notion of dream as content.[18] This is figured, in the dream, by the *spaced out* formula for the chemical 'Trimethylamin', a representation of a different mode of reading via typography, to which I will return. Freud, then, recognizes the necessity of transference and proposes a complex theory of reading which accounts for the plurality and mobility of signs, but remains monolithic and linear in his approach. There is a parallel between psychoanalysis and the detective novel in that both propose complex models of reading complex textual surfaces – but in each this theory of reading is countered by a version of the truth, by a search for truth which remains monolithic and repressive of this complexity.

What is the limit, then, of the reading of signs? What does Poe tell us about reading? The detective is a reader of material signs, clues, as the psychoanalyst is a reader of the material of language. But what kind of sign is a clue? The detective Dupin, or Holmes, only gains access to the truth through a particular attention to the material quality of the signifier. The clue is a sign which has been diverted from its everyday use and which has as a result taken on a particular material, textural or textual quality which only the detective can spot. The detective is an adherent of a radically different materialistic discourse which reverses the Platonic philosophy whereby forms represent ideas. The system of signs on which culture and commerce are based is overturned, becomes a mere materiality, but one which can be diverted into profit for the detective. In this reversal, the object is taken away from its usual signification; it no longer functions as a simple receptacle for an idea but is seen in (as opposed to being seen *as*) its actuality, its density. It is re-endowed with physicality and corporeality. Thus, in 'The Murders in the Rue Morgue', when the police have failed, Dupin reveals the explanation of the crime: how the murderer gets out of the locked room hinges on the simple material fact of a nail whose head has broken off. This technique is in every sense parallel to the *ostranie*, or 'making strange' effect which Victor Schlovsky attributes to poetry.[19] Dupin of course is also a poet. Perhaps it is this quality which makes him a better reader of material signs than Freud. But this is still not enough, for as Derrida points out in his critique of Lacan's seminar, the 'Purloined Letter' is part of a trilogy, completed by 'The Murders in the Rue Morgue' and the 'Mystery of Marie Roget'.[20] They are meant to be understood together, for if Dupin solves the Rue Morgue mystery by recognizing that the solution to such an extraordinary murder must lie in an *extraordinary* means, the escaped orang-utang, through the insight that the witnesses heard the culprit speaking an unidentified language *which was not French*, he finds the Purloined Letter through the opposite ruse: the recognition that the mystery is too plain for the police to solve, that the letter is not hidden but disguised as itself, in its everyday use.

The police, and the psychoanalyst, fail because they always look for the hidden, the secret. Detective fiction is often fetishistic in its approach to reading. It focuses on the object in order to find the truth behind it, and the greater its focus on the object the greater its desire or fear of the thing which is hidden behind it. Rather than the pleasure with materiality of a Dupin or a Holmes, the dull detective (or contemporary reader) is fixed and fascinated in front of the object as fetish, suspecting a traumatic truth behind it. Dupin's method holds a lesson for a culture obsessed with the hidden, yet it is not an endorsement for all that of the emasculation of meaning executed by postmodernity. Engagement with a different mode of reading, one deriving from material, textual pleasure, from the texture of experience, is not the same as the forced imposition of violent or simply banal experience which is that of the postmodern.

But the best detective fiction is not just about the search for the hidden truth. The truth is disappointment, the relapse of syntax into a linear monovalence. Poe's 'The Mystery of Marie Roget' breaks off before even getting there. Subverting this linearity is an indulgence in metanarrative, in intertextuality. Poe's story is made up predominantly of Dupin's reading of newspaper reports. Chandler's heroes take pleasure in masochistic alienation from the corrupt mass and indulge in what Freud would call 'short circuits' of repression, jokes, far more than they apply themselves to the search for truth. The pleasure to be got out of detection, or reading, is not only the gratification of that infantile search for the one hidden grail or phallus, *objet petit a*, which lies hidden somewhere, but the perverse, baroque pleasure of material, textual surfaces. Poe's manuscript found in a bottle, already a typographic figure, includes a figure of textuality not determined by intention, and arising, as it were, out of the folds of matter, when the narrator unwittingly daubs with a tar-brush the edges of a neatly folded studding sail which lies near him on a barrel. He writes: 'The studding-sail is now bent upon the ship, and the thoughtless touches of the brush are spread out into the word DISCOVERY.'[21] The notion that truth and destiny arise out of and are somehow encrypted in the fold, as the purloined letter is folded, proposes a radically different version of the truth from the linear and monolithic, phallogocentric version of psychoanalysis. The truth is nothing, a radical negation, a hole: the hole toward which this ship is speeding, the maelstrom,[22] Mr Valdemar's wordless voice from beyond the grave.[23]

Two possible paths open up here: the link of the detective novel to the psychoanalytic vision of death, which might look closely at the 'compulsion to repeat' in the service of the death drive, in relation to the clue, or to the detective's relation to the corpse. The other path is to return to what Freud represses, according to Mehlman, in the dream of Irma's injection, but what he radically proposes despite himself with the figures of displacement and condensation: the dream as *syntax*, or the dominance of the syntactic over the semantic, of texture over truth. Dupin insists also on the game of puzzles which requires the player to find a hidden word on the 'perplexed surface' of a chart.[24] The superior player selects 'such words as stretch, in large characters, from one end of the chart to the other. These, like the over-largely lettered signs and placards of the street, escape observation by dint of being excessively obvious; and here the physical oversight is precisely analogous with the moral inapprehension by which the intellect suffers to pass unnoticed those considerations which are too obtrusively and too palpably self-evident'.[25] The 'palpably self-evident' is the evidence which can be touched, which is there to be read in its material texture, which holds the truth in the multiple folds of its perplexed surface.

If Poe's stories show an ambiguous tension between the detective as master and the differential and multiple quality of truth, then the detective genre which his three stories apparently inaugurate (though he abandoned this mode afterwards), and psychoanalysis too, show themselves up to be also characterised by a dynamic between their own textuality and their will for mastery over the 'perplexed surface' of experience and textuality; in this will for mastery on the part of detective and analyst, both display the symptoms of an *infantile* quality, a will to control the idiosyncrasies of the text. Perhaps this infantile quality has made them particularly vulnerable to a culture which celebrates the superficial and emasculates meaning to the profit of the enjoyment of affective violence. One might propose as potential antidotes, however, those misnamed 'postmodern' narratives which offer their readers an indulgent enjoyment of complexity, which are seemingly indifferent to the truth of the crime. One could point to the writings of Pynchon, Auster or Robbe-Grillet as texts which frustrate, if one is looking for the hidden truth, or texts which reward, if one takes pleasure in the complexity of the fold. In terms of the representation of trauma, in the light of my reading, Perec's *La Disparition* would show a trauma not hidden *behind* but only too evident on the surface of the text: the only too explicit absence of the letter *e* figuring the *disappearance* of *eux* (them), and thus inscribing the trauma of the holocaust on the surface of the text.

If the parallel between the detective and the psychoanalyst reveals a complicity in a fetishistic anxiety about a hidden truth, the paradigm, in Derridean terms, of this anxiety and this complicity would be the resolute belief in a transcendent God, a postulation of a presence beyond or behind the material texture of things. To continue the somewhat projective, or wishful, character of much of what I have advanced here, I would like to end with the utopian proposition, which counters this anxiety, of a materialist detective, analyst, or reader whose pleasure is invested in what Freud calls perversion, a 'lingering' on the pathway to full sexual fulfilment, on the way to discovery of the hidden truth.

Notes

1 'Led by some obscure feeling [lay opinion] seems to assume that, in spite of everything, every dream has a meaning, though a hidden one, and dreams are designed to take the place of some other process of thought and that we have only to undo the substitution correctly in order to arrive at this hidden meaning.' Sigmund Freud, *The Interpretation of Dreams* (Harmondsworth: Penguin, 1976), p. 169.
2 Stephen Heath, 'Notes on Suture', in *Screen,* 18.4 (Winter 1977–8), 48–76.
3 Jacques Derrida sees the expression 'la psychanalyse se trouve' as curiously symptomatic of psychoanalysis: it finds itself in everything which it considers and particularly in the 'heterogeneous and conflictual weave of *différance*'. Jacques Derrida, *The Postcard: From Socrates to Freud and Beyond* (Chicago and London: University of Chicago Press, 1987), p. 413.
4 Edgar Allan Poe, *Tales of Mystery and Imagination* (London: Dent, 1981), pp. 378–9.
5 Marie Bonaparte, *Edgar Poe, sa vie, son œuvre* (Paris: PUF, 1933), trans. John Rodker as *The Life and Works of Edgar Allan Poe: A Psycho-analytical Interpretation* (London: Hogarth Press, 1949).
6 Jacques Lacan, 'Seminar on the Purloined Letter', trans. J. Mehlman in *Yale French Studies,* 48 (1973), 39–72.
7 Jacques Derrida, 'Le facteur de la vérité', in *The Postcard.*
8 See for example, John P. Muller and William J. Richardson (eds), *The Purloined Poe: Lacan, Derrida & Psychoanalytic Reading* (Baltimore: Johns Hopkins University Press, 1988).
9 Poe, *Tales of Mystery and Imagination,* pp. 379–80.
10 Cf. Poe, 'The Mystery of Marie Roget', author's note: '*The Mystery of Marie Roget* was composed at a distance from the scene of the atrocity, and with no other means of investigation than the newspapers afforded', *Tales of Mystery and Imagination,* p. 410.
11 Julia Kristeva, 'Le sens et la mode', in *Séméiotiké* (Paris: Seuil, 1969).
12 For an account of this shift, see Patrick ffrench, *The Time of Theory: A History of* Tel Quel (Oxford: Clarendon, 1996).
13 Freud, *The Interpretation of Dreams,* pp. 170–1.
14 Ibid., p. 173.
15 Ibid., p. 178.
16 Derrida, *The Postcard,* p. 414.
17 Ibid., p. 422 and *passim.*
18 Jeffrey Mehlman, 'Trimethylamin: Notes on Freud's Specimen Dream', in Robert Young (ed.), *Untying the Text: A Post-Structuralist Reader* (London: Routledge, 1981).
19 Cf. Victor Schlovsky, 'Art as Process', in Victor Erlich, *Russian Formalism: History, Doctrine* (The Hague: Mouton, 1965).

20 Derrida, *The Postcard*, pp. 458–9.
21 Poe, 'Manuscript found in a Bottle', in *Tales of Mystery and Imagination*, p. 264.
22 Cf. Poe, 'Descent into the Maëlstrom', in *Tales of Mystery and Imagination*, pp. 243–58.
23 Cf. Poe, 'The Facts in the Case of M. Valdemar', in *Tales of Mystery and Imagination*, pp. 280–9.
24 Poe, 'The Purloined Letter', in *Tales of Mystery and Imagination*, p. 467.
25 Ibid., p. 468.

Index